Petrissa Eckle

Attosecond Angular Streaking

Petrissa Eckle

Attosecond Angular Streaking

Attosecond time resolution with femtosecond laser pulses

Südwestdeutscher Verlag für Hochschulschriften

Impressum/Imprint (nur für Deutschland/ only for Germany)
Bibliografische Information der Deutschen Nationalbibliothek: Die Deutsche Nationalbibliothek
verzeichnet diese Publikation in der Deutschen Nationalbibliografie; detaillierte bibliografische
Daten sind im Internet über http://dnb.d-nb.de abrufbar.
Alle in diesem Buch genannten Marken und Produktnamen unterliegen warenzeichen-, marken-
oder patentrechtlichem Schutz bzw. sind Warenzeichen oder eingetragene Warenzeichen der
jeweiligen Inhaber. Die Wiedergabe von Marken, Produktnamen, Gebrauchsnamen,
Handelsnamen, Warenbezeichnungen u.s.w. in diesem Werk berechtigt auch ohne besondere
Kennzeichnung nicht zu der Annahme, dass solche Namen im Sinne der Warenzeichen- und
Markenschutzgesetzgebung als frei zu betrachten wären und daher von jedermann benutzt
werden dürften.

Verlag: Südwestdeutscher Verlag für Hochschulschriften Aktiengesellschaft & Co. KG
Dudweiler Landstr. 99, 66123 Saarbrücken, Deutschland
Telefon +49 681 37 20 271-1, Telefax +49 681 37 20 271-0, Email: info@svh-verlag.de
Zugl.: Zürich, ETH Zurich, Diss.,2008

Herstellung in Deutschland:
Schaltungsdienst Lange o.H.G., Berlin
Books on Demand GmbH, Norderstedt
Reha GmbH, Saarbrücken
Amazon Distribution GmbH, Leipzig
ISBN: 978-3-8381-0685-4

Imprint (only for USA, GB)
Bibliographic information published by the Deutsche Nationalbibliothek: The Deutsche
Nationalbibliothek lists this publication in the Deutsche Nationalbibliografie; detailed
bibliographic data are available in the Internet at http://dnb.d-nb.de.
Any brand names and product names mentioned in this book are subject to trademark, brand or
patent protection and are trademarks or registered trademarks of their respective holders. The
use of brand names, product names, common names, trade names, product descriptions etc.
even without a particular marking in this works is in no way to be construed to mean that such
names may be regarded as unrestricted in respect of trademark and brand protection legislation
and could thus be used by anyone.

Publisher:
Südwestdeutscher Verlag für Hochschulschriften Aktiengesellschaft & Co. KG
Dudweiler Landstr. 99, 66123 Saarbrücken, Germany
Phone +49 681 37 20 271-1, Fax +49 681 37 20 271-0, Email: info@svh-verlag.de

Copyright © 2009 by the author and Südwestdeutscher Verlag für Hochschulschriften
Aktiengesellschaft & Co. KG and licensors
All rights reserved. Saarbrücken 2009

Printed in the U.S.A.
Printed in the U.K. by (see last page)
ISBN: 978-3-8381-0685-4

Table of Contents

Table of Contents I
List of Figures III
Publications VII
    Journal Papers VII
    Conference Papers VII
Abstract IX
Kurzfassung (German) XI
Introduction 1
Time in Tunneling 5
    2.1   Introduction 5
    2.2   The Wigner-Eisenbud-Smith time or phase time 6
    2.3   Buttiker-Landauer Time 7
    2.4   The adiabaticity parameter of Keldysh 10
    2.5   A tunneling delay time in high field ionization 12
Attosecond Angular Streaking (AAS) 15
    3.1   Introduction 15
    3.2   The concept of attosecond angular streaking 16
    3.3   The pulse field 18
        3.3.1   The Carrier Envelope Offset Phase 18
        3.3.2   CEP in circular polarization 19
        3.3.3   Ellipticity effects 21
        3.3.4   Absolute value of the CEP in circular and elliptical light 23
    3.4   Streaking 24
    3.5   Comparison to energy streaking 27
Semi-classical simulations 29
    4.1   Introduction 29
    4.2   Pulse simulation 29
        4.2.1   Phase and spectrum 30
        4.2.2   Quarter wave plate 31
    4.3   Streaking in the field 33
    4.4   Ionization, ADK rates 34
    4.5   Intensity calibration 35
Experimental setup I: The Laser Pulse 37
    5.1   Laser 37
    5.2   Pulse compression by filamentation 40
    5.3   Pulse characterization: Spider 42
    5.4   CEP stabilization 46
Experimental setup II: COLTRIMS 49
    6.1   Introduction 49
    6.2   Gas Jet 50
        6.2.1   Spectrometer 51
    6.3   Detection 52
    6.4   Signal processing 53
    6.5   Calibration 53
CEP measurement with AAS 57
    7.1   Experimental setup 58

7.2	Data	59
7.3	Resolution and accuracy	61
7.4	Simulation	61
7.5	Data analysis	62
	7.5.1 Temporal accuracy	62
	7.5.2 Temporal resolution	63
	7.5.3 CEP resolution	64

Absolute time: Tunneling ionization dynamics **67**

8.1	Introduction	67
	8.1.1 Relative and absolute measurement of the tunneling delay time	67
	8.1.2 Ellipticity dependence	69
	8.1.3 CEP dependence	70
8.2	Experimental details	71
	8.2.1 Setup	71
	8.2.2 Angular calibration	72
	8.2.3 Measurement procedure	73
8.3	Measurement of the tunneling delay time in an ellipticity scan	74
	8.3.1 Polarization scan	74
	8.3.2 Ion measurement	76
8.4	Data analysis	77
	8.4.1 Simulations	77
	8.4.2 Streaking angle and tunneling delay time	77
	8.4.3 Statistical error limit	78
8.5	Intensity dependence of the tunneling delay time	79
	8.5.1 Intensity scans	80
	8.5.2 Experiment	81
8.6	Conclusion	83

Summary & Outlook **84**

References **88**

Acknowledgements **95**

List of Figures

Figure 2-1 Tunneling through a time dependent rectangular barrier. The tunneling particle can lose or pick up energy quanta of the barrier oscillation with different probability   7

Figure 2-2 a) Multiphoton ionization, the electron absorbs multiple photons and thereby more energy than is needed for ionization b) tunneling ionization c) Over barrier ionization, the potential is bent so far that the elctron becomes free   11

Figure 2-3 The upper panel shows a linearly polarized few cycle pulse. The maximum electric field defines $t_{0,Field}$. Below the corresponding ionization rate is shown with the maximum rate defined as $t_{0,ion}$. In an experiment, $t_{0,Field}$ of the electric field is usually not accessible and therefore assumed to be equal to $t_{0,ion}$ of the corresponding ionization rate.   13

Figure 3-1 The two steps in AAS: Ionization and streaking.   17

Figure 3-2 Left panel: The 'atto-clock': mapping time to momentum. Right panel: red solid: electric pulse field of a circularly polarized 6 fs pulse, blue dotted: corresponding ionization rate with FWHM ~2 fs.   17

Figure 3-3 left:CEP in a linearly polarized ultrashort pulse for CEP = 0 (red solid) and $\pi/2$ (blue broken line). Right: Intensity as a function of time of the pulse on the left. Again for CEP = 0 (red solid) and $\pi/2$ (blue broken line).   19

Figure 3-4: Circularly polarized pulse in space and time (upper panel) and the temporal evolution (lower panel). In b, the CEP is shifted by $\pi/4$ compared to a, rotating the pulse in space by $\pi/4$ without changing the shape of the temporal evolution.   21

Figure 3-5 The polarization ellipse is shown in black, broken line. On the left, the spatial evolution of an elliptically polarized few-cycle pulse in the polarization plane is shown, where the envelope maximum of the electric field points into the direction of the major axis of the electric field. On the right the same pulse is shown but oriented along the minor axis of the polarization ellipse.   22

Figure 3-6: CEP dependence of the electric field depicted in Figure 3-5 in elliptically polarized light. The black broken line shows the temporal envelope of a circularly polarized pulse. On the left (red solid line), the temporal field evolution is shown for the envelope pointing into the direction of the major axis of the ellipse, corresponding to a CEP=0, on the right (blue, solid line), the envelope points along the minor axis of the ellipse, CEP=$\pi/2$.   23

Figure 3-7 Shown is the evolution of an elliptically polarized pulse with a field envelope wit FWHM =3.7 fs in the polarization plane for different ellipticities. In blue, solid line the pulse's electric field is depicted. The polarization ellipse is oriented vertically (90 degrees), the electric field envelope points along the minor axis of the ellipse. Black dotted lines indicate the pointing of the electric field maxima. The ellipticities from left to right are: ε =0.8, 0.9, 0.98 and 1, i.e. circular, respectively.   24

Figure 3-8 Left: Streaking angle for a Gaussian pulse for perfectly circular light (blue solid line) and an ellipticity of 0.048 (green dotted). On the right: corresponding ionization rates.   26

Figure 3-9 Adapted from [55]. The initial electron distribution is mapped depending on the timing between the attosecond pulse and the streaking field.   27

Figure 4-1 The refractive index of quartzglass in the wavelengt range around 800 nm   31

Figure 4-2 Data of the $\lambda/4$ plate. Phase, angle of the principal axis and ellipticity as a function of wavelength. The dotted line in the first panel indicates the $\pi/4$ phase shift required for perfectly circularly polarized light.   33

Figure 4-3 Ionization rates for circularly polarized pulses calculated with ADK   35

Figure 4-4 Ionization rate as a function of time for the same pulse field but different intensities of an elliptically polarized pulse calculated with ADK formulas. The rates are normalized to peak rates. The dotted line corresponds to a peak intensity of $1\cdot 10^{15}\,W/cm^2$, the dashed line to $5\cdot 10^{14}\,W/cm^2$ and the solid line to $1\cdot 10^{14}\,W/cm^2$. 36

Figure 5-1 The essentially transform limited oscillator pulse is stretched by chirping it, then amplified and recompressed. 38

Figure 5-2 The laser system consisting of the fs-oscillator, amplifier and prism compressor. Pump light is shown as green, thick lines, the IR pulse is shown in red, thin lines. Only four passes through the amplifier crystal are visible since the pulse travel out of the plane. 40

Figure 5-3 Spectrum and phase of a 34 fs pulse (left) and the temporal intensity envelope (right) 40

Figure 5-4 The two mechanisms that form a dynamic equilibrium leading to filamentation 41

Figure 5-5 The two stage filamentation setup. The gas cells were filled with 650-900 mbar of argon. Chirped mirrors were used for compression. 42

Figure 5-6 Left: Spectrum (solid line) and phase (broken line) of a 5 fs pulse. Right: The temporal intensity envelope 42

Figure 5-7 SPIDER setup: Short pulses are depicted as dotted lines, the stretched pulse is depicted as a solid line. 44

Figure 5-8 f-to-2f scheme for determining the CEO frequency of a pulse train. 47

Figure 5-9 Measured (black trace) and reconstructed (gray trace) single-shot CEO spectral interference pattern. The reconstructed pattern was calculated using the phase(ψ (ω)) and amplitude information obtained from the Fourier filtering phase reconstruction technique. 48

Figure 6-1 Coltrims setup showing the gas jet, laser and spectrometer with detectors for ions and electrons 49

Figure 6-2 Raman mapping of rotational temperatures in a supersonic jet of CO_2 under a stagnation pressure of 2 bars. Isothermal lines are depicted at steps of 20 K. 50

Figure 6-3 Detector with MCP (right) and delay line anode (left) 52

Figure 6-4 the left panel shows electron data. The distribution in the x-direction of the detector is plotted versus the time of flight of the electrons, clearly showing the refocusing in time caused by the magnetic field of the detector. On the right some electron trajectories are plotted, in light blue the refocusing times are marked. 54

Figure 7-1 The upper panel shows the CEP values over measurement time for each single shot as a gray level encoded histogram, darker colors correspond to a higher density of measured CEP values. The center of the distribution is the CEP value, the width of the distribution is the uncertainty in CEP . The lower panel shows the corresponding error signal for the feedback, slowly shifting the phase shot by shot. The full range of the feedback signal is 4095 steps, in the first 30 minutes the error signal was almost constant meaning that the system was drifting. 59

Figure 7-2 Overview over the measured helium ion momentum distributions while scanning the CEP over 2π and comparison with a semi classical simulation. The top row shows measured momentum distribution for four different values of the CEP. In the lower panel the momentum distributions are radially integrated and their angular dependence on the CEP is shown for the full scan of the CEP over 2π for both data (on the left) and simulation (on the right). 60

Figure 7-3 CEP dependence of the ionization angle in helium using attosecond angular streaking: a, b: Radially integrated ion momentum distributions for two values of the CEP, the ellipticity peaks are fitted with double Gaussians to extract $\theta_{1,2}$ c, d: The angular position of

the two peaks θ_1 and θ_2 as a function of the CEP with simulations indicated as a dashed line.
63

Figure 8-1 Left: the polarization ellipse in red, solid line. Center: the ionization rate ellipse depending on the CEP. On the right: The rate ellipse is streaked in the pulse field and rotated by approximately 90 degrees. 68

Figure 8-2 shows the strong dependence of the streaking angle on the ellipticity of the pulse. The streaking angle is simulated using the measured pulse parameters and the calculated broadband wave plate used in the experiment: The green dotted line gives the ellipticity as a function of wave plate angle, the black solid line shows the corresponding streaking angle. The streaking angle shows the strongest variation around 45 degrees, where the light is closest to circular. 69

Figure 8-3 Shown are the ionization rate distributions of a few cycle elliptically polarized pulse in the polarization plane for two different CEP values as a solid line. The broken line shows the orientation of the polarization ellipse. On the left, the distribution is shown for a CEP of $\pi/2$, on the right for a CEP of $-\pi/2$. If the two distributions are averaged, the resulting maxima are aligned with the polarization ellipse. 71

Figure 8-4 The upper panel shows the setup for the polarization measurement, in the lower panel polarizer and power meter are removed to measure the corresponding ion momentum distributions. 72

Figure 8-5 Calibration with linear light. On the left the ion momentum distribution in the plane perpendicular to the laser propagation is shown. Vertically polarized light along the y-direction creates a cigar-shaped momentum distributions along the y-direction. On the right the polarization measurement and fit to the radially integrated ion momentum distribution from the right panel is shown, since this measurement is used to calibrate the coordinate system between ion measurements and polarization measurements the angle $\Delta\theta$ is set to zero.
73

Figure 8-6 On the left, a typical ion momentum distribution projected onto the polarization plane is shown. On the right as in Figure 8-5, in the upper panel the polarizer scan is shown with the angular orientation of the polarization ellipse, θ_{field}. The lower panel shows the corresponding angular distribution of the ions with the maximum θ_{ions} shifted by the streaking angle $\Delta\theta$ 74

Figure 8-7 Measurement of θ_{field}. Scan of the wave plate angle and thus the ellipticity. Shown as dots is the measured orientation of the polarization ellipse, the solid line shows the simulated orientation. The green squar dots show the ellipticity extracted fom the measurement and the dotted line gives the corresponding simulated ellipticity curve. 75

Figure 8-8 Measurement of θ_{ions} Scan of the wave plate angle and thus the ellipticity. Shown as dots is the measured orientation of the ellipticity peaks on the ion momentum distribution, the dashed line shows the simulated orientation. 76

Figure 8-9 The streaking angle $\Delta\theta$ from the measurement is shown as dots with error bars derived from the individual measurements of θ_{field} and θ_{ions}. The black broken line shows the corresponding simulation. The calculated values were corrected by the effect of the coulomb potential yielding the values shown as a black solid line. The green dotted line gives the corresponding ellipticity indicated on the axis on the right. 78

Figure 8-10 Measured tunneling time delay $\Delta\tau_D$ (data points) for a fixed ellipticity (i.e. 0.88) as a function of peak intensity and Keldysh parameter. The tunneling time delay is the difference between the measured and the calculated streaking angle assuming instantaneous tunneling. In the calculated streaking angle we also took into account the Coulomb potential (solid line set to zero). The intensity-averaged offset is 6.0 as with a standard deviation of 5.6 as. The two dimensional helium momentum distributions measured at different intensities and the corresponding radially integrated distributions spanning one optical cycle (2.44 fs) are shown on the right. The time axis is calibrated and the calculated streaking time is deducted. 82

Publications

Parts of this thesis have been published in the following journal papers and conference proceedings:

Journal Papers

C. P. Hauri, A. Guandalini, P. Eckle, W. Kornelis, J. Biegert, U. Keller, *"Generation of intense few-cycle laser pulses through filamentation – parameter dependence"*, Opt. Express, vol. 13, No. 19, pp. 7541-7547, 2005

A. Guandalini, P. Eckle, M. P. Anscombe, P. Schlup, J. Biegert, U. Keller, *"5.1 fs pulses generated by filamentation and carrier envelope phase stability analysis"*, J. Phys. B: At. Mol. Opt. Phys., vol. 39, pp. S257-S264, 2006

P. Eckle, M. Smolarski, P. Schlup, J. Biegert, A. Staudte, M. Schöffler, H. G. Muller, R. Dörner, and U. Keller, *"Attosecond Angular Streaking,"* Nat. Phys., vol. 4, pp. 565-570, 2008.

P. Eckle, A. Pfeiffer, C. Cirelli, A. Staudte, R. Dörner, H. G. Muller, M. Büttiker, and U. Keller, *"Attosecond ionization and tunneling delay time measurements"*, Science, Science, Vol. 322. no. 5907, pp. 1525 – 1529, 2008

E. Mansten, J. M. Dahlström, J. Mauritsson, T. Ruchon and A. L'Huillier
J. Tate and M. B. Gaarde, P. Eckle, A. Guandalini, M. Holler, F. Schapper, L. Gallmann, and U. Keller, *"Spectral signature of short attosecond pulse trains"*, Phys. Rev. Lett., accepted

Conference Papers

P. Schlup, P. R. Eckle, A. Aghajani-Talesh, J. Biegert, M. P. Smolarski, A. Staudte, M. Schöffler, O. Jagutzki, R. Dörner, U. Keller, *"Ionization of Ar with Circularly Polarized 5.5-fs Pulses for the Determination of CEO Phase"*, Conference on Lasers and Electro-Optics (CLEO'06), Long Beach, California, USA, May 21-26, 2006

A. Guandalini, P. Eckle, M. Anscombe, P. Schlup, J. Biegert and U. Keller, *"Intense CEO phase stabilized ultra-short pulses approaching the single cycle limit"*, Conference on Lasers and Electro-Optics (CLEO '06), Long Beach, California, USA, May 21-26, 2006

A. Guandalini, P. Eckle, F. Schapper, A. Couairon, M. Franco, A. Mysyrowicz, J. Biegert, U. Keller, *"5.1-fs pulses by filamentation – prospective of self-compression to one optical cycle"*, 15th International Conference on Ultrafast Phenomena, Pacific Grove, California, USA, July 31-Aug. 4, 2006

P. R. Eckle, P. Schlup, J. Biegert, M. P. Smolarski, A. Staudte, M. Schöffler, O. Jagutzki, R. Dörner, U. Keller, *"Determination of the CEO phase – ionization of He with circularly polarized*

5.5-fs pulses", 15th International Conference on Ultrafast Phenomena, Pacific Grove, California, USA, July 31- Aug. 4, 2006

J. Biegert, P. Eckle, P. Schlup, M. Smolarski, A. Staudte, M. Schöffler, O.Jagutzki, R.Dörner, U. Keller, *"Subcycle spatial mapping of ionization dynamics in He"*, Super Intense Laser Atome Physics (SILAP 2006), Salamanca, Spain, June 19-23, 2006

J. Biegert, P. R. Eckle, P. Schlup, U. Keller, M. P Smolarski, A. Staudte, M. Schöffler, O. Jagutzki, R. Dörner, *"Spatial mapping of sub-cycle dynamics"*, International Laser Physics Workshop (LPHYS'06), Lausanne, Switzerland, July 24-28, 2006

P. Eckle, M. Smolarski, P. Schlup, J. Biegert, A. Staudte, M. Schöffler, H. G. Muller, R. Dörner, U. Keller, *"Attosecond angular streaking"*, Ultrafast Optics and High Field Short Wavelength Sources ((UFO/HFSW'07), Santa Fe, USA, Sep. 2-7, 2007

C. Cirelli, P. Eckle, M. Smolarski, A. Pfeiffer, P. Schlup, J. Biegert, A. Staudte, M. Schöffler, H.G. Muller, R. Dörner, U. Keller, *"Attosecond angular streaking"*, Attosecond Physics (Atto'07), Int. Workshop and 391. WE-Heraeus Seminar, Dresden, Germany, August 2007

P. Eckle, M. Smolarski, A. Staudte, M. Schöffler, P. Schlup, J. Biegert, H.G. Muller, R. Dörner, U. Keller, *"Attosecond angular streaking"*, SPS Annual meeting, Geneva, Switzerland, Mai 2008

P. Eckle, A. Pfeiffer, C. Cirelli, U. Keller, R. Dörner, A. Staudte, H. G. Muller, M. Büttiker, *"Attosecond angular streaking: an ideal technique to measure electron tunneling time?"*, Conference on Ultrafast Phenomena, Stresa, Italy June 9- 13, 2008

A. N. Pfeiffer, P. Eckle, C. Cirelli, A. Staudte, R. Dörner, H. G. Muller, M. Büttiker, U. Keller, *"Attosecond Ionization and Tunneling Delay Time Measurements"*, 11th International Conference on Multiphoton Processes (ICOMP'08), Heidelberg, Germany, Sep. 18–23, 2008

Abstract

In this thesis a new technique called 'attosecond angular streaking' (AAS) was applied for the first time. AAS allows to resolve ionization dynamics in the strong field regime with attosecond accuracy ($1 as = 10^{-18} s$) using only femtosecond pulses ($1 fs = 10^{-15} s$). In this regime, ionization mainly proceeds via tunneling through an energetically forbidden barrier.

The process of tunneling is a fundamental and well-understood phenomenon in quantum mechanics. However, attempts to measure the 'tunneling time' have produced controversial results, partly due to the difficulty of defining a temporal operator in quantum mechanics, making it difficult to compare experiments.

In this thesis, the question is addressed of whether the tunneling rate can adjust instantaneously to a changing barrier or if there is a delay $\Delta \tau_D$ between the field that defines the barrier and the corresponding tunneling ionization rate. To measure this delay required not only the temporal resolution of the ionization rate but also an independent measurement of the temporal evolution of the electric field with sub-cycle accuracy.

State of the art laser pulses with durations between five to ten femtoseconds easily reach peak electric field strengths comparable to those that bind electrons in an atom. This so-called strong field regime opened the door to phenomena such as high harmonic generation that finally led to the generation of attosecond pulses, to above threshold ionization and tunneling ionization through the potential barrier suppressed by the pulse field. In this regime the interaction of the laser pulse with an atom depends highly nonlinearly on the instantaneous electric field strength of the pulse. The field goes through one oscillation in around 2.5 femtoseconds so that temporal dynamics proceed on a sub-femtosecond timescale.

AAS is used to resolve these sub-cycle dynamics in time. As soon as an electron is set free through ionization it is accelerated under the influence of the rotating electric field vector of a circularly polarized few-cycle pulse and deflected in its angle. The direction of its final momentum after the pulse has passed then is a direct measure of the instant of ionization: AAS maps time to angle.

In a proof of principle experiment the carrier envelope phase (CEP) of the circularly polarized pulse was measured, demonstrating a temporal accuracy of 23 as and a resolution around 200 as.

To study the temporal dynamics of the tunneling ionization process AAS then was applied in combination with a reference measurement of the electric field.

In a static picture the tunneling rate is highest when the barrier is lowest. In the experiment presented here, the circularly polarized pulse modulates the tunneling barrier with the carrier frequency of the pulse so that width and height of the barrier change while the particle is crossing it. This raises the question if the tunneling process leads to a delay between the change of the barrier and the corresponding tunneling rate that could then be interpreted as a 'tunneling delay time'.

The key to measure the time delay between the electric field and the corresponding tunneling rate was to find a 'marker' that is measurable in the field as well as in the ionization rate giving a reference time for both measurements. A slight ellipticity provides this marker by modulating the electric field strength on a sub-cycle time scale. With this method, $\Delta\tau_D$ was measured for a range of ellipticities and for various intensities. It was found that indeed, no delay between the electric field and the corresponding tunneling rate exists within an experimental error limit of 12 as.

Kurzfassung (German)

In dieser Arbeit wurde attosecond angular streaking (AAS), eine neue Technik, zum ersten Mal angewandt. AAS erlaubt es, unter Verwendung von lediglich Femtosekundenpulsen ($1 fs = 10^{-15} s$), zeitliche Ionisationsdynamiken im Bereich starker Felder im Attosekundenbereich ($1 as = 10^{-18} s$) aufzulösen. Der dominante Ionisationsmechanismus in diesem Starkfeldbereich ist Tunneln durch eine energetisch verbotene Barriere.

Der Tunnelprozess ist ein fundamentales und gut verstandenes quantenmechanischen Phänomen. Versuche, die Tunnelzeit zu messen haben

jedoch zu kontroversen Ergebnissen geführt. Dies ist zum Teil durch die Schwierigkeit bedingt, eine quantenmechanische Zeitobservable zu definieren, was die Vergleichbarkeit von Messungen erschwert.

Die vorliegende Arbeit beschäftigt sich mit der Frage, ob die Tunnelrate sich instantan an eine sich ändernde Barriere anpassen kann oder ob es eine Zeitverzögerung $\Delta \tau_D$ zwischen dem Feld, das die Barriere bestimmt und der dazugehörigen Tunnelionisationsrate gibt. Um diese mögliche Zeitverzögerung zu messen war nicht nur die zeitliche Auflösung der Ionisationsrate notwendig, sondern auch eine unabhängige Messung der zeitlichen Entwicklung des elektrischen Feldes mit einer weit höheren Genauigkeit als einer der Schwingungsperiode des Pulses (sub-cycle).

Laserpulse auf dem aktuellen Stand der Technik mit Pulsdauern zwischen 5 und 10 Femtosekunden erreichen leicht elektrische Feldstärken, die mit den Feldstärken, die Elektronen im Atom binden, vergleichbar sind. Dieser so- genannte Stark-Feld-Bereich eröffnete den Zugang zu Phänomenen wie der Erzeugung höherer Harmonischer, die schliesslich zur Erzeugung von Attosekundenpulsen führte, zur „Above Threshold Ionization" (ATI) und zur Tunnelionisation durch eine Barriere, die durch das Feld des Pulses unterdrückt wird. In diesem Bereich hängt die Wechselwirkung zwischen dem Laserpuls und einem Atom hoch nichtlinear von der momentanen Feldstärke des Pulses ab.

Das elektrische Feld hat eine Periode von ungefähr 2.5 fs, so dass sich die zeitliche Dynamik auf einer sub-Femtosekunden Zeitskala abspielt.

AAS wird verwendet, um diese sub-cycle Dynamik in der Zeit aufzulösen. Sobald ein Elektron durch Ionisation freigesetzt ist, wird es unter dem Einfluss des rotierenden elektrischen Feldvektors eines zirkular polarisierten ultrakurzen Pulses beschleunigt und in seinem Winkel abgelenkt. Die Richtung seines letztendlichen Impulses, nachdem der Laserpuls vorüber ist, ist dann eine direktes Mass für den Ionisationszeitpunkt: AAS bildet Zeit auf einen Winkel ab.

In einem ersten Experiment zur Verifikation der Technik wurde die absolute Phase (Carrier Envelope Offset Phase - CEP) des zirkular polarisierten Pulses gemessen; Dabei ergaben sich eine zeitliche Genauigkeit von 23 as und eine Auflösung von etwa 200 as.

AAS wurde dann in Kombination mit einer Referenzmessung des elektrischen Feldes angewandt, um die zeitliche Dynamik des Tunnelionisationsprozesses zu untersuchen.

In einem zeitunabhängigen Bild ist die Tunnelrate am höchsten, wenn die Barriere am niedrigsten ist. In dem hier gezeigten Experiment moduliert der zirkular polarisierte Puls die Tunnelbarriere mit der Zentralfrequenz des Pulses, so dass Breite und Höhe der Barriere sich ändern, während das Teilchen die Barriere durchquert. Dies wirft die Frage auf, ob der Tunnelprozess zu einer Verzögerung zwischen der Änderung der Barriere und der dazugehörigen Tunnelrate führt, die dann als ‚Tunnel-Verzögerungszeit' interpretiert werden könnte.

Um diese Zeitverzögerung zwischen dem elektrischen Feld und der zugehörigen Tunnelrate zu messen galt es, als Schüssel einen Marker zu finden, der sowohl im elektrischen Feld als auch in der Ionisationsrate messbar ist und so eine zeitliche Referenz für beide Messungen gibt. Ein solcher Marker kann durch kleine Elliptizität erzeugt werden, die die elektrische Feldstärke auf einer sub-cycle Zeitskala moduliert. Mit dieser Methode wurde $\Delta\tau_D$ für verschiedene Elliptizitäten und Intensitäten gemessen. Das Ergebnis war, dass mit einem oberen Limit von 12 as, bedingt durch die Messungenauigkeit tatsächlich keine Verzögerung zwischen der maximalen Feldstärke und der zugehörigen maximalen Tunnelrate besteht.

Chapter 1

Introduction

An attosecond is the billionth of a billionth of a second; it is to one second, as one second is to the age of the universe. But despite these 18 orders of magnitude, time resolution on an attosecond scale has come within reach of today's experiments.

The human eye can resolve time down to around a twentieth of a second. This is not even sufficient to clearly see our own macroscopic world. In sports events for example, players are captured with high-speed cameras and the sequence is played back slowly. Much the same approach is possible also in the microscopic world, providing there is a fast enough camera.

Just as short illumination with a flashlight allows to freeze motion and shot by shot make a movie out of a succession of images, lasers have made it possible to continue making movies of nature to ever-faster times scales.

In chemical reactions the temporal dynamics are governed by the speed of the nuclei, their translation, rotation and vibration. One important question is thus, how energy is distributed over different degrees of freedom and how fast this happens. The nuclei in motion define the potential surfaces on which electrons proceed to new steady states forming or destroying bonds between atoms. The fastest molecular dynamics proceed on a femtosecond timescale.

In traditional chemistry only the starting and the final products of a chemical reaction are known, revealing very little about intermediate products or states of the molecules involved. Femtosecond chemistry allowed for the first time to detect which electronic excitations play a role in a reaction, how molecules change shape in the course of a reaction, and how molecules interact with their environment, e.g. a solution. The latter topic is particularly important to understand the chemistry of living organisms, where reactions take place in an environment of water.

In 1999 the Nobel Prize was awarded to Ahmed Zewail [1] for 'filming' for the first time how a chemical bond breaks, the fundamental process of a chemical reaction.

Laser pulses allow thus to initiate and manipulate processes dynamically [2]. Unlike the analogy with filming suggests, laser pulses do not merely watch the temporal evolution, they interact with the systems that are investigated.

Laser pulses allow also to concentrate high energy in a short time interval, strongly perturbing the molecular structure. This opens up reaction pathways that are hidden in traditional thermally driven chemistry and new effects emerge, such as above threshold dissociation, bond-softening, bond hardening and vibrational trapping [3].

The fastest molecular motion is the fundamental vibrational mode of the smallest molecule, a hydrogen molecule, where one oscillation lasts around 5 fs. (check number)

The internal timescale of systems is related to their size (or more precisely, the masses and forces involved). Smaller systems show faster dynamics: Looking deeper into the atom to electronic dynamics changes the timescale from femto- to attoseconds. For electron dynamics, the fundamental timescale is given by the time the electron of a hydrogen atom needs to complete one roundtrip around its proton, namely 24 attoseconds.

Attoseconds are by no means the limit [4]. The next step to a smaller system is the atomic nucleus itself, where processes are expected to proceed in the order of $10^{-21} s$, at one thousandth of an attosecond. The final limit is far away. It is thought to be the so-called Planck-time, where time should become quantized, at $10^{-43} s$.

Since the laser was invented in 1954, dramatic progress has been made decreasing pulse lengths. During the last years femtosecond pulses have become ever shorter [5, 6]. Pulses based on visible to infrared frequencies are approaching the border of a single cycle of the field oscillation. In the case of the commonly used carrier wavelength of 800 nm one oscillation takes 2.7 fs.

To overcome this border and to advance truly into the regime of attosecond physics, the central wavelength of pulses needed to become shorter. Pulses were shifted to the XUV, significantly raising the technological demands for generation and control. Through high harmonic generation in noble gas targets, pulses as short as 100 and 80 as [7, 8] could be generated, but applying these pulses in a measurement remains a challenge. Pulses have to be handled entirely under vacuum, since air is strongly absorbing in the XUV. Bandwidths of these pulses are very broad, making it difficult to design mirrors. Nevertheless, attosecond pulses have been very successfully applied for time resolved measurements.

Attosecond pulses were used to directly resolve the electric field oscillation of a 5 fs pulse in time [9]. The first application of an attosecond pulse to measure a physical process in an atom was to confirm the duration of an important multi electron process, the Auger decay [10]: If an electron is removed from an inner shell leaving a so called core hole, an electron from a higher shell can take its place and shed its excess energy either through emission of a photon or by transferring it to a third, the so called Auger electron that leaves the atom. The duration of this process can be deduced from measurements in the energy domain by analyzing the line shape; the detailed dynamics however remain hidden. In principle the durations for such Auger processes range from attoseconds to a few femtoseconds. In this case, the duration was measured with an attosecond pump pulse creating the core hole and a femtosecond probe to 7.9 fs.

At this stage, attosecond pulses cannot be used as just a shorter replacement for the femtosecond pulses used before. One reason is that they cannot be produced with enough intensity to induce strong field processes. The photon energies of attosecond pulses range from 10 to hundreds of eV and are thus much higher compared to the around 1.5 eV at 800 nm central wavelength. The intensities of these pulses on the other hand are reach up to $10^9 W/cm^2$ and thus are orders of magnitudes smaller than the current typical intensities of femto-IR pulses around $10^{15} W/cm^2$. The physical mechanisms through which these pulses interact with atoms can thus not be compared to infrared femtosecond pulses. For the low intensity XUV pulses, a single photon picture is appropriate and nonlinear effects play no role. IR pulses on the other hand reach easily intensities where the electric field becomes comparable to the binding energy of valence shell electrons. At these intensities ionization cannot be described any more by the absorption of single photons, the light looses its quantum nature, instead the pulse can be treated as a classical electric field bending the potential of the atom.

This strong field regime allows to study a fundamentally quantum mechanical process, tunneling through an energetically forbidden barrier. Electrons can escape their atom from a bound state by crossing the lowered potential barrier of the Coulomb force without having the energy to surmount it. While this process is well known, the question for its duration has remained a controversial issue (for details see Chapter 2).

To investigate these tunneling dynamics, both strong fields and attosecond temporal resolution are necessary.

One possibility is again to combine an attosecond pulse as a trigger and then induce the tunneling process with a strong femtosecond pulse. This experiment was performed by exciting an atom with a 250 as long XUV pulse. The attosecond pulse thus acted as the trigger to time the beginning of the tunneling process. The excited atom was subsequently tunnel ionized by the field of a 5 fs intense infrared pulse [11].

In this thesis a different approach will be followed to investigate tunneling dynamics with attosecond resolution. A new technique called attosecond angular streaking (AAS) is introduced in Chapter 3 that allows to achieve attosecond resolution and accuracy without actually using attosecond pulses. Instead the electric field of a circularly polarized pulse of around 5.5 fs duration is used to map temporal dynamics to momentum space, a parameter that is experimentally much better accessible. This technique is first used to measure the precise waveform of the electric field of the pulse. This experiment presented in Chapter 6 demonstrates that an accuracy in the temporal measurement of sub-100 as can be achieved.

In Chapter 7, attosecond angular streaking is applied to measure the tunneling time delay that was defined in chapter 2.5.

Crucial for the temporal accuracy and resolution that can be achieved with AAS are precise simulations, presented in Chapter 4. Details of the experimental setup are given in Chapter 5 and Chapter 6.

Chapter 2

Time in Tunneling

2.1 Introduction

Quantum tunneling remains one of the most puzzling processes predicted by quantum mechanics. Even if it is acceptable that a particle passes through a potential barrier if it does not have the energy to surmount it, trouble starts when asking about the time the particle actually took to tunnel. This question has been discussed ever since the tunneling effect was discovered [12] and has created a wealth of literature covered in several review papers [13-15]. A historic overview on the topic of tunneling time can be found in [16].

The question 'how long does it take for a particle to tunnel' looks rather innocent, but it provokes controversy on multiple levels, ultimately touching on the unique role of time itself in quantum mechanics. In the Schrödinger equation, time is a parameter, while measurable physical quantities, observables, are linked to operators that have to fulfill the formal condition of being self-adjoint to yield real values. Pauli noted [17] that such an operator could only exist for systems with continuous energy spectrum, recent research however seems to contradict this [18-20].

Even though it might not be possible to find a universal time operator, it was shown that operators can be constructed to yield meaningful specific time-observables such as the arrival time or dwell/sojourn time (the time a particle stays in a confined area of space that contains a barrier in a scattering event) [15, 21].

Specific times that have all been used as a synonym for tunneling time are the 'dwell-time', 'delay-time', 'reflection-time' and 'traversal-time' [22], one reason why there is such confusion around the topic.

Recent efforts mainly center around uniting and reconciling the different definitions of 'tunneling times', e.g. by developing an 'umbrella theory' [23]. Here, to calculate the time a particle spent in the barrier, two operators D and P are defined. D chooses the part of the particle's wave function that is found inside a barrier region limited by [a,b]. P chooses the part of the wave function that is actually transmitted. These operators are shown not to commute, so that it matters which selection is made first. An infinite number of combinations of these operators is possible, meaning that there are many different possibilities to define a tunneling time. This of course implies that in tunneling time measurements the question must be very carefully phrased to find meaningful definitions of tunneling times that are at the same time experimentally accessible.

The analysis of some simple cases derived in this umbrella theory is shown to yield several possible definitions of a tunneling time that have already been discussed in the literature, notably covering two important approaches that will be discussed in more detail in chapter 2.2 and 2.3, respectively, the Wigner-Eisenbud-Smith time [24, 25] or phase time and the Buttiker-Landauer time [26], that is also called semi-classical tunneling time in the literature.

Chapter 2.5 presents another tunneling time: If an atom is exposed to the oscillating electric field of a laser pulse that suppresses its atomic potential, an initially bound electron can tunnel through the barrier. The tunneling rate depends on the strength of the suppressing field and the question arises if the maximal tunneling rate is reached at the peaks of the electric field or if there is delay between the field and the resulting tunneling rate. This possible delay is experimentally investigated in this thesis in Chapter 8.

2.2 The Wigner-Eisenbud-Smith time or phase time

Historically the most important approach to tunneling time was probably the Wigner-Eisenbud-Smith time, not least because it comes closest to the classical picture of a particle whose timing, position and speed are measured before and after a barrier.

The Wigner-Eisenbud-Smith approach considers a wave packet, e.g. a laser pulse or the wave function of a particle and follows its peak. The derivative of the transmission phase shift (hence also the name phase time) then yields the group velocity v_g of the wave packet. This approach leads to the famous paradox that a particle tunneling through a barrier traverses the same distance actually faster than a particle passing through free space. Furthermore the tunneling speed can even become superluminal [27], the so called Hartmann effect. The tunneling time becomes independent of the barrier thickness d for wide barriers with very low transmission probability, so called opaque barriers. If the tunneling velocity is then defined as $v_g = d / \tau_{Ph}$ it is obvious that there is no limit to the tunneling speed.

Many experiments on varying systems have confirmed, that these seemingly paradoxical effects are measurable [28-35]. Common agreement today is however, that this is no violation of causality. It was pointed out, that there is no conservation law in physics for peaks of wave packets [26]. The answer of Sommerfeld and Brillouin to superluminal velocities in regions of anomalous dispersion was to characterize pulse propagation by a signal velocity, which is always limited by the velocity of light [36, 37], see also [38]. Recently a different interpretation was given by [39], where the group delay is interpreted as a cavity life time of the particle's wave function in the barrier rather than a traversal time through the barrier.

2.3 Buttiker-Landauer Time

The approach of Büttiker and Landauer [26] avoids measuring a wave packet peak; instead, the barrier itself is used to probe how long the particle took to tunnel so that the time is encoded in the particle's energy spectrum. Other than the phase time this approach always yields subluminal tunneling velocities.

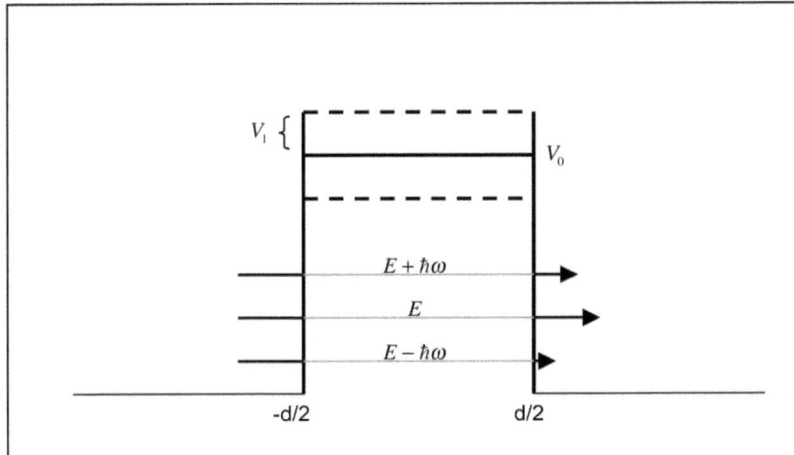

Figure 2-1 Tunneling through a time dependent rectangular barrier. The tunneling particle can lose or pick up energy quanta of the barrier oscillation with different probability

In the approach of Büttiker and Landauer the barrier through which the particle tunnels is not stationary but oscillates in height as shown in Figure 2-1. The frequency of this oscillation is used to probe the duration of the interaction between the tunneling particle and the barrier.

This model system is particularly interesting in the frame of this thesis because it corresponds closely to the experimental situation of a single atom where the atomic potential is periodically modulated by the electric field of a strong laser pulse.

For a rectangular oscillating barrier, the time dependent potential can be written as:

$$V(x) = V_0(x) + V_1(x)\cos(\omega t)$$

There are two limiting cases. In the non-adiabatic regime the period of the oscillation is much shorter than the time the particle interacts with the barrier, so it feels a time independent, averaged potential. In the adiabatic case, if the period of the modulation is much longer than the interaction time, the particle sees a stationary barrier of varying height. The Buttiker-Landauer tunneling time is then defined by the crossover frequency ω_c between these two limiting cases, where the period of the modulation equals the interaction time of the particle with the barrier:

$$\tau_{BL} = 1/\omega_c$$

The interaction time for the rectangular barrier shown in Figure 2-1 can be shown to yield (eq (2) in [26]):

$$\tau_{BL} = \int_{-d/2}^{d/2} \sqrt{\frac{m}{2[V_0(x)-E]}} dx, \quad E < V_0,$$

where m is the particle's mass and E its initial energy.

The key to experimentally access the crossover frequency and thus the tunneling time is the energy spectrum of the tunneled particles.

During its interaction with the barrier, the particle can change its initial kinetic energy E by absorbing or emitting energy quanta of the barrier oscillation $\hbar\omega$. These sidebands of $\pm\hbar\omega$ in the energy of the transmitted particles yield the crossover frequency ω_c:

In the low-frequency limit, where the tunneling process proceeds much faster than the barrier oscillation, the sidebands are equally strong, while for the case of a fast oscillating barrier the transmission amplitude T_+ for particles that pick up one quantum of energy is enhanced compared to the amplitude of the first low energy sideband with amplitude T_-. This can be intuitively understood as a higher energy particle traverses the barrier more easily. The barrier interaction time can then be expressed through an asymmetry parameter of the transmission amplitudes of the first two sidebands:

$$\frac{(T_+ - T_-)}{(T_+ + T_-)} = \tanh(\omega \tau_{BL})$$

The suggestion of the Büttiker-Landauer time as characteristic timescale of the average interaction time was followed by several experimental implementations in slightly modified model systems [40], the most important being the following two experiments:

In one experiment [41, 42] an electron is tunneling through a barrier in a semiconductor micro structure and during the tunneling feels the attraction of its electrostatic image charge on the material it left. The formation of the image charge through polarization has a characteristic response time, i.e. the plasma frequency of the electrons in the material, setting the crossover frequency against which the tunneling time can be measured. In this experiment it is confirmed that for thinner barriers that show shorter tunneling times than the response time, the influence of the image charge is negligible, while it becomes significant when the tunneling time becomes comparable or longer than the response time.

The key idea of the second experiment that confirmed the findings of Büttiker and Landauer was to send a feedback triggered by the tunneling process back to the system while tunneling is not completed. If the feedback is delayed, its influence on the tunneling process should successively decay. The half-life of this decay can then be viewed as the crossover time τ_{BL}.

In this experiment the tunneling of an entire electrical circuit containing a Josephson junction from a metastable state was investigated [40, 43].

A Josephson junction is essentially a resistor sandwiched between two superconductors. Particles (cooper pairs of electrons in this case) can tunnel through the resistor, creating a current that travels along a variable delay line that closes the circuit acting as a feedback signal. This feedback hinders the tunneling process and lengthens the lifetime of the metastable state. With longer and longer delays the feedback looses influence until the lifetime saturates. The crossover time i.e. the Büttiker-Landauer was defined as the half-life of the decay function of the feedback strength. The macroscopic nature of this process leads to a comparably very long tunneling time of 78 ps, but still no quantitative comparison with τ_{BL} was possible since the system could not be modeled accurately enough.

Defining a tunneling time via a crossover frequency thus yields an experimentally accessible quantity; in the case of the experiment investigating the spreading of the image charge it proved to be a useful quantity to predict tunneling rates in semiconductor structures.

2.4 The adiabaticity parameter of Keldysh

The model system used to investigate tunneling time in this thesis is an atom under the influence of the electrical field of an intense laser pulse. The pulse field that lowers the potential barrier of the atom varies on the same time scale as the expected tunneling time in the Büttiker-Landauer approach.

To characterize tunneling in this strong field regime, Keldysh introduced the dimensionless adiabaticity parameter γ [44], that is closely related to the Buttiker-Landauer time: $\gamma = \tau_{BL}\omega_0$. It is used in high-field physics to distinguish between the regimes where the ionization rate follows the field adiabatically and the regime where ionization proceeds on a much larger timescale than the oscillation of the laser field that perturbs the atomic potential. It is defined as

$$\gamma = \frac{\omega_0\sqrt{2I_p}}{E_0},$$

where I_p is the ionization potential, ω_0 is the angular frequency of the laser and E_0 its maximum electric field amplitude (all in atomic units).

For 800 nm radiation the energy of a single photon in the pulse field is 1.55 eV and so by far smaller than the ionization potential of most atoms and small molecules. This means that ionization by single photons plays no role. At sufficiently high intensities ionization becomes nevertheless possible through nonlinear effects.

For $\gamma \gg 1$, i.e. short wavelength and low intensity, the ionization is most properly described by the simultaneous absorption of many discrete photons. (See Figure 2-2). This mechanism is also called multiphoton ionization (MPI) or above threshold ionization (ATI) since usually more photons are absorbed than necessary to surmount the ionization threshold.

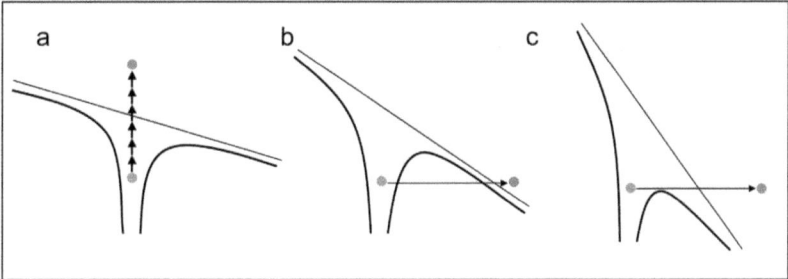

Figure 2-2 a) Multiphoton ionization, the electron absorbs multiple photons and thereby more energy than is needed for ionization b) tunneling ionization c) Over barrier ionization, the potential is bent so far that the elctron becomes free

At higher intensities the electron can tunnel through into the continuum. On the other hand, if $\gamma \ll 1$, the electric field strength becomes similar to the binding energy of the outermost electron, the potential of the atom is considerably lowered by the electric field and tunneling of the bound electron through a classical potential barrier provides the appropriate physical picture of the ionization process. The two regimes are not expected to have a sharp transition and generally both tunneling and MPI are assumed to be present in the intermediate regime with $\gamma \sim 1$. This regime is referred to as the regime of non-adiabatic tunneling [45]. Keldysh parameters for typical experimental values are calculated in Table 2-1.

wavelength	Intensity	γ linear polarization	γ circular polarization
600 nm	$1 \cdot 10^{14} W / cm^2$	1.91	2.71
600 nm	$4 \cdot 10^{14} W / cm^2$	0.96	1.35
600 nm	$8 \cdot 10^{14} W / cm^2$	0.68	0.96
800 nm	$1 \cdot 10^{14} W / cm^2$	1.01	1.43
800 nm	$4 \cdot 10^{14} W / cm^2$	0.72	1.02
800 nm	$8 \cdot 10^{14} W / cm^2$	0.51	0.72
1000 nm	$1 \cdot 10^{14} W / cm^2$	1.15	1.62
1000 nm	$4 \cdot 10^{14} W / cm^2$	0.57	0.81
1000 nm	$8 \cdot 10^{14} W / cm^2$	0.41	0.57

Table 2-1 shows Keldysh parameters for different wavelength and different intensities for Helium, ionization potential 24.58 eV

Finally, for even higher intensity the potential barrier is completely suppressed by the electric field and the electron is set free by over barrier ionization .

Note that for the same peak intensity the Keldysh parameter γ is $\sqrt{2}$-times larger for circularly polarized light than for linearly polarized light.

While it has been possible to develop rate equations for each of the regimes separately, the intermediate regime where $\gamma \sim 1$ remains both theoretically challenging and experimentally interesting.

2.5 A tunneling delay time in high field ionization

Experiments on high field ionization with femtosecond or even attosecond resolution have brought up another tunneling time, called delay time τ_D in the tunneling process [11, 46].

Most experiments in high field physics take place in the regime around a Keldysh parameter of $\gamma = 1$, the regime of non-adiabatic tunneling [45] (see chapter 2.4). It was shown that even at $\gamma \approx 3$ the ionization rate traces the oscillations of the electric field [11] that suppresses the potential barrier of an atom. Figure 2-3 shows the field of a linearly polarized 5 fs pulse and the corresponding ionization rate calculated with ADK formulas (see chapter 4.4).

However, usually only the time dependence of the ionization rate is experimentally accessible and it remains unclear, if the instant of the maximum suppressing electric field ($t_{0,field}$) corresponds to the instant of maximum ionization rate ($t_{0,ion}$).

Recent experiments have addressed tunneling time in strong field ionization and an upper limit to the tunneling time could be established by triggering the tunneling process with an attosecond pulse.

In [11] the temporal evolution, tunneling from excited state of a neon ion (Ne^{+*}) was measured. Neon atoms were first ionized into an excited state with an attosecond pulse and subsequently tunnel ionized to Ne^{2+}. The attosecond pulse confines the ionization-excitation event and thus the start of the tunneling process to around 250 as. The yield of Ne^{2+} ions then depends on the instantaneous intensity of the electric field of the IR pulse that enables the tunneling process by suppressing the potential of the Ne^{+*}. Measuring the delay between the attosecond trigger event and the IR probe then gives an indication of the timescale of the tunneling process. One issue in this case is that only the combined time of excitation and tunneling can be measured. The other issue is that the absolute value of the delay between the attosecond pulse and the streaking field can not be measured, i.e. the question if the ionization rate adjusts adiabatically to the variation in barrier height or if there is any delay between the suppressing field and the resulting ionization rate remains unanswered. In this experiment there was no way to access this information independently from the described measurement and it was assumed that the instant where the most particles are set free into the continuum corresponds to the highest instantaneous laser field.

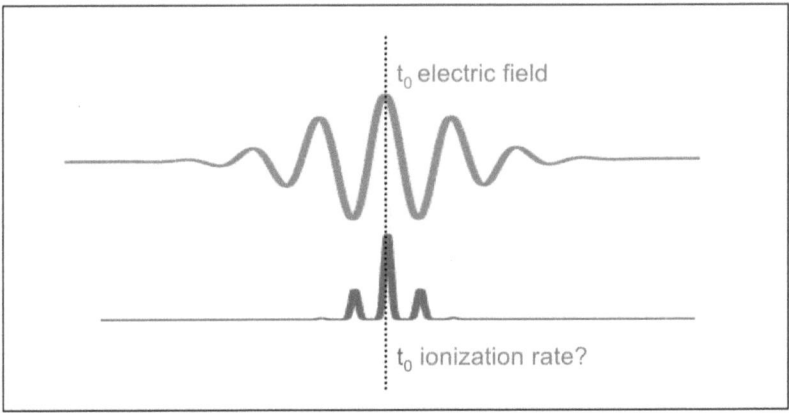

Figure 2-3 The upper panel shows a linearly polarized few cycle pulse. The maximum electric field defines $t_{0,Field}$. Below the corresponding ionization rate is shown with the maximum rate defined as $t_{0,ion}$. In an experiment, $t_{0,Field}$ of the electric field is usually not accessible and therefore assumed to be equal to $t_{0,ion}$ of the corresponding ionization rate.

The key to answering the question if $t_{0,field}$ equals $t_{0,ion}$ is an independent measurement of the electric field with attosecond accuracy. This is not a trivial task, and for linear light has not been achieved so far. In the experiments presented in this thesis it will be shown that with elliptically polarized light a 'timing marker' can be introduced that is measurable in the field as well as in the ionization rate.

Chapter 8.1 explains how an ellipticity in the electric pulse field can be used as such a marker, the results of the measurement of τ_D are presented in Chapter 8.

Chapter 3

Attosecond Angular Streaking (AAS)

3.1 Introduction

In this thesis the temporal dynamics of a single atom in an intense laser pulse is investigated. Two timescales govern the temporal evolution in this system:

An external timescale is set by the oscillation period of the laser, i.e. 2.4 fs. The internal timescale on which the system evolves is determined by the roundtrip time of the electrons in their shells, for the electron of a hydrogen atom this yields the atomic unit of time, 24 attoseconds.

Direct time measurements converting time to electronic signals can offer at best picosecond resolution, but indirect time domain measurement techniques such as pump probe and streaking that were developed to go beyond the barrier set by the resolution of electronic devices could successfully be implemented up to the attosecond regime [47].

In pump probe measurements, a short pulse triggers a process and a second pulse arriving at a variable delay is used to probe the dynamics. Pump-probe is thus a sampling technique requiring a repeatable process and the temporal resolution is limited by the length of the pulses.

Streaking techniques circumvent both these limitations by projecting a temporal evolution e.g. to space or momentum space, variables that are usually experimentally much better accessible than time. Since the projecting function establishes a unique relation between time and the measurement variable, streaking can be used as a single shot measurement. Several ways to apply streaking in the attosecond regime were proposed [48] using femtosecond laser pulses as streaking field. Energy streaking, i.e. mapping time to energy, was used to measure the electric field of a linearly polarized 5 fs pulse with an attosecond trigger pulse [9].

The technique presented in this thesis uses circularly polarized light to map time to angular momentum, it was named attosecond angular streaking (AAS) [49]. Different implementations of AAS are possible (see chapter 3.5), in all the experiments presented in this thesis however, a ~5.9 fs circularly polarized laser pulse was used for both ionization and as streaking field. In the experiment presented in this thesis, without the use of attosecond pulses a temporal resolution of ~200 as and a temporal accuracy of ~24 as could be achieved.

3.2 The concept of attosecond angular streaking

In AAS, time is mapped to angular momentum using the rotating electric field vector of a circularly polarized laser pulse as streaking field. Depending on the time of ionization electrons are deflected in the angular spatial direction, so that the instant of ionization is mapped to the final angle of the momentum vector in the polarization plane. The 'atto-clock' runs over one 360-degree turn of the electric field in 2.4 fs. It requires pulse durations in the two optical cycle regime.

The process of AAS can be divided into two steps, the ionization and the streaking. This two step model is analogous to the three step model that successfully explained high harmonic generation [50], where it was shown that the ionization process and the subsequent acceleration of the electron in the electric field of the laser pulse can be treated independently, neglecting further Coulomb interaction between the electron and its parent ion during the motion in the laser field.

The two steps are shown in Figure 3-1. First the particle tunnels through the potential barrier of the atom or molecule that is suppressed by the pulse's electric field. It appears in the continuum with essentially zero momentum and is accelerated by the rotating electric field vector of the circularly polarized laser in radial and angular direction.

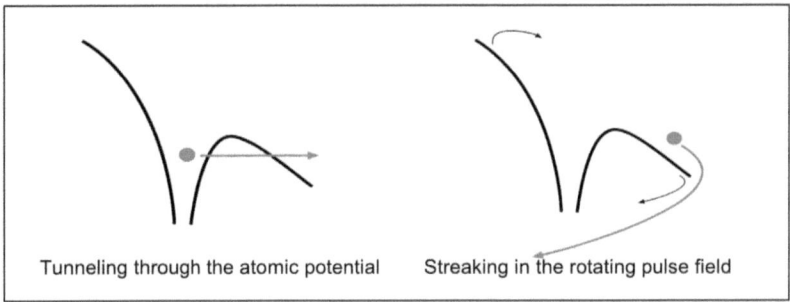

Figure 3-1 The two steps in AAS: Ionization and streaking.

Other than for linear light, where particles are accelerated and decelerated, in circular polarization continuous gain of momentum leads to a toroidal distribution evolving around the laser axis. The particles start into the direction of the field vector and are deflected from their original direction by around 90 degrees ahead of the field at the time of ionization as shown in Figure 3-2 (see also chapter 3.4).

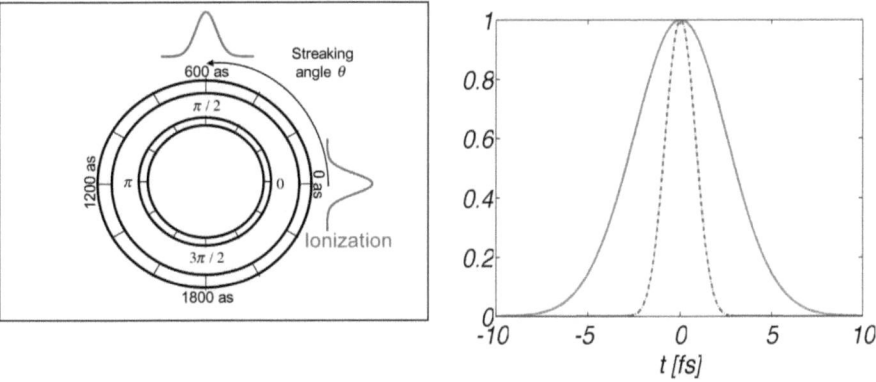

Figure 3-2 Left panel: The 'atto-clock': mapping time to momentum. Right panel: red solid: electric pulse field of a circularly polarized 6 fs pulse, blue dotted: corresponding ionization rate with FWHM ~2 fs.

This means that the final momentum of the particle is a direct mapping of the instant it was born into the continuum. In the polarization plane, the momentum distribution forms the 'atto-clock'.

To ensure that the momentum distribution is a unique mapping of ionization times, the time span during which ionization takes place needs to be sufficiently short (i.e. shorter than one cycle of the streaking field), a condition that is fulfilled for the ~6 fs pulse used in the experiment, due to the high nonlinearity of the ionization probability in the laser field. This is shown in Figure 3-2 in the right panel: The red solid line shows the intensity envelope of a circularly polarized pulse, the blue dotted line shows the corresponding ionization rate with a FWHM of ~ 2 fs. For such short pulses, the particles also move only around five nanometers during the duration of the pulse, so that they do not leave the focal volume of the laser or even see a field gradient in the spatial direction, significantly simplifying calculations.

3.3 The pulse field

Both ionization and streaking depend sensitively on the electric field of the pulse, requiring precise control and analysis of the pulse field to exploit the full potential of the AAS.

In strong-field ionization, the ionization rate depends highly nonlinearly on the instantaneous electric field strength of the pulse, so that even small features in the pulse field are strongly enhanced. The streaking angle on the other hand depends linearly on the electric field and the mapping function can be shown to be strictly monotonic. Nevertheless, the total streaking angle varies by several degrees during the pulse, depending on the exact waveform of the streaking field.

3.3.1 The Carrier Envelope Offset Phase

In a few cycle pulse an important parameter that determines the shape of the electric field of the pulse is the carrier envelope offset phase (CEP) [51-53].

In general, the temporal evolution of an electric field of a linearly polarized pulse can be split into the fast oscillation of the carrier with frequency ω_0 and the envelope $E_0(t)$:

$$E(t) = E_0(t)\cos(\omega_0 t + \phi_{CEO}),$$

At a wavelength of 725 nm, the duration of one laser oscillation is 2.4 fs. This means that at pulse durations around 6 fs the pulse envelope changes almost as fast as the electric field oscillation, so the shape of the electric field of the pulse becomes significantly dependent on ϕ_{CEO}, the CEP, that sets the relative position of the carrier wave and the envelope.

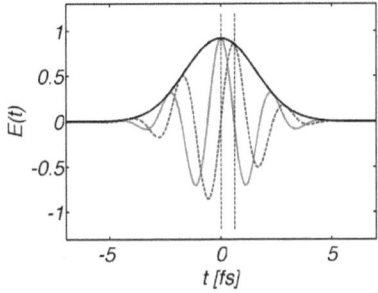 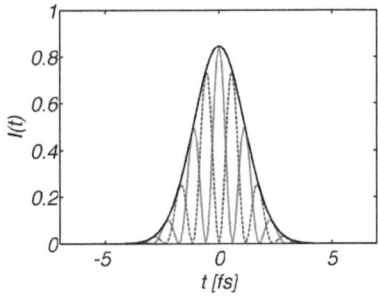

Figure 3-3 left:CEP in a linearly polarized ultrashort pulse for CEP = 0 (red solid) and $\pi/2$ (blue broken line). Right: Intensity as a function of time of the pulse on the left. Again for CEP = 0 (red solid) and $\pi/2$ (blue broken line).

In Figure 3-3 on the left, the shape of the electric field of a ~2 cycle pulse is depicted for two different values of the CEP. A so called 'cosine pulse' is shown in red, solid line, where the maximum of the oscillation of the carrier frequency coincides with the maximum of the pulse envelope. The intensity of the pulse depicted on the right $I(t) \sim |E(t)|^2$ then shows one distinct maximum. In blue (broken line) a 'sine pulse' is shown, where a minimum in the carrier field coincides with the maximum of the pulse envelope. The intensity then has two equally strong peaks.

During propagation in any medium the CEP changes continuously due to the difference between the phase velocity that determines the speed of the carrier wave and the group velocity of the envelope. In a freely running laser the CEP fluctuates randomly from pulse to pulse, it is, however, possible to actively stabilize the phase (see chapter 5.4).

3.3.2 CEP in circular polarization

The red line in the upper panel of Figure 3-4 shows circularly polarized pulses in three dimensions, the polarization plane x-y and time. The field maximum E_{max} is indicated by a dot.

The electric field of a circularly polarized pulse can be written as a sum of two orthogonal components with a relative phase difference of $\pi/2$. The two components are shown in the upper panel of Figure 3-4 as projections onto the x-t and the y-t plane respectively. They can be written as follows:

$$E_x(t) = E_0(t)\sin(\omega_0 t + \phi_{CEO})$$

$$E_y(t) = E_0(t)\cos(\omega_0 t + \phi_{CEO})$$

with equal temporal pulse envelopes

$$E_0(t) = E_0 e^{-a \cdot t^2}, \quad a = \frac{4\ln 2}{FWHM^2}$$

where FWHM is the full-width at half the maximum of the electric field envelope.

In circular polarization, the electric field vector rotates in the polarization plane with the carrier frequency, changing its magnitude smoothly with the pulse envelope.

The time dependence of the electric field vector reads:

$$|E(t)| = \sqrt{E_x(t)^2 + E_y(t)^2} = E_0(t) \cdot \sqrt{\left[\cos(\omega_0 t + \phi_{CEO})\right]^2 + \left[\sin(\omega_0 t + \phi_{CEO})\right]^2} = E_0(t)$$

This shows, that in perfectly circularly polarized light, the temporal evolution of the pulse field is independent of the CEP. In the lower panel of Figure 3-4, the corresponding electric field of the two pulses in a and b is shown in time: The oscillation of the carrier frequency is not visible in the temporal domain.

Instead the CEP determines the orientation of the pulse in space, which becomes immediately clear if the pulse field is written in polar coordinates (r,ϕ) projected onto the x-y plane:

$$r(t) = |E(t)|$$

$$\phi(t) = \arctan\left(\frac{E_0(t)\sin(\omega_0 t + \phi_{CEO})}{E_0(t)\cos(\omega_0 t + \phi_{CEO})}\right) = \arctan\left[\tan(\omega_0 t + \phi_{CEO})\right] = (\omega_0 t + \phi_{CEO})$$

The field maximum is reached for $t=0$, so that $\phi_{max} = \phi_{CEO}$. If the CEP of the two components is shifted simultaneously by $\Delta\phi$, the pulse field is rotated in space by $\Delta\phi$. This is shown in Figure 3-4: the CEP in **b** is shifted by $\pi/4$ compared to **a**, rotating the pulse by 45 degrees.

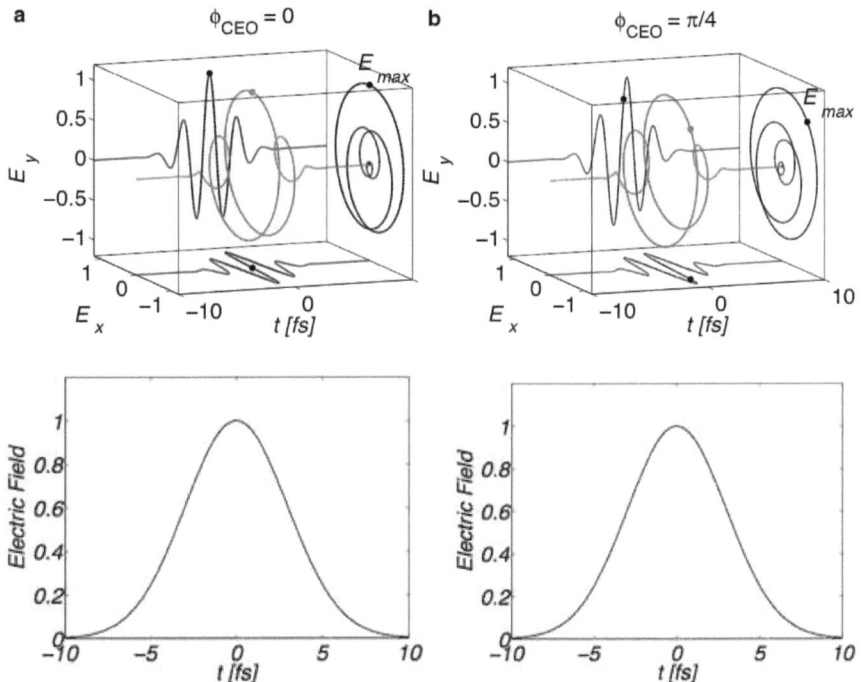

Figure 3-4: Circularly polarized pulse in space and time (upper panel) and the temporal evolution (lower panel). In b, the CEP is shifted by $\pi/4$ compared to a, rotating the pulse in space by $\pi/4$ without changing the shape of the temporal evolution.

3.3.3 Ellipticity effects

In elliptically polarized light, the magnitude of the two field components is not equal any more. The field components then read:

$$E_x(t) = E_0(t)\cos(\omega_0 t + \phi_{CEO})$$

$$E_y(t) = \varepsilon E_0(t)\sin(\omega_0 t + \phi_{CEO})$$

where $\varepsilon = \left[\dfrac{E_y(t)}{E_x(t)}\right]$ is the ellipticity of the pulse.

In elliptically polarized light, the magnitude of the electric field vector traces the polarization ellipse going trough maxima when it is oriented along the major axis of the ellipse and minima along the minor axis. The evolution in the x-y plane is depicted in Figure 3-5.

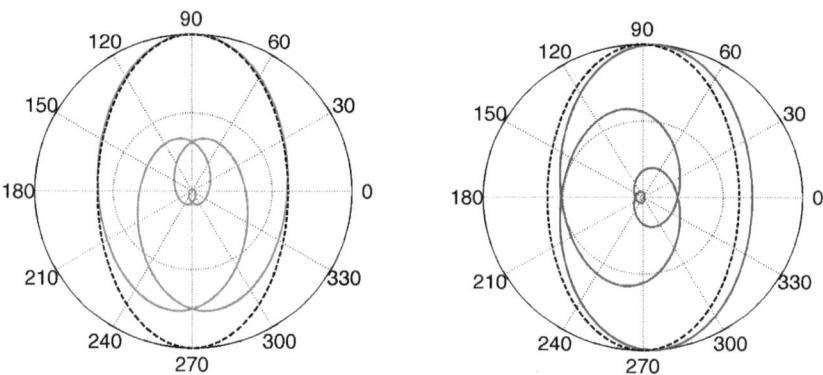

Figure 3-5 The polarization ellipse is shown in black, broken line. On the left, the spatial evolution of an elliptically polarized few-cycle pulse in the polarization plane is shown, where the envelope maximum of the electric field points into the direction of the major axis of the electric field. On the right the same pulse is shown but oriented along the minor axis of the polarization ellipse.

Compared to the circular case, the spatial evolution of the pulse is elongated along the major axis of the ellipse.

Analogous to circular light, the magnitude is calculated to be:

$$|E(t)| = \sqrt{E_x(t)^2 + E_y(t)^2} = E_0(t) \cdot \sqrt{\left[\cos(\omega_0 t + \phi_{CEO})\right]^2 + \varepsilon^2 \left[\sin(\omega_0 t + \phi_{CEO})\right]^2}$$

This can be rewritten as:

$$E_0(t) \cdot \sqrt{1 + (\varepsilon^2 - 1)\left[\sin(\omega_0 t + \phi_{CEO})\right]^2}$$

The temporal evolution thus can be split into the original envelope $E_0(t)$ and an additional oscillation with twice the carrier frequency of the original pulse field and amplitude ε. This is shown in Figure 3-6. The black broken line shows the original pulse envelope. The solid lines in both graphs shows the envelope with the oscillations stemming from the tracing of the polarization ellipse on top. When the envelope is multiplied with the oscillation, 'ellipticity peaks' appear in the field. The temporal field shows one distinct maximum if the envelope points along the direction of the major axis of the pulse (Figure 3-6, left). If the CEP of the pulse is changed the pulse is rotated in space while the polarization ellipse remains fixed, so that in the temporal evolution of the field the ellipticity peaks are shifted. If the phase is rotated by $\Delta\phi_{CEO} = \pi/2$ and thus points into the direction of the minor axis of the ellipse this yields two equal ellipticity peaks on the pulse envelope as shown in Figure 3-6, right.

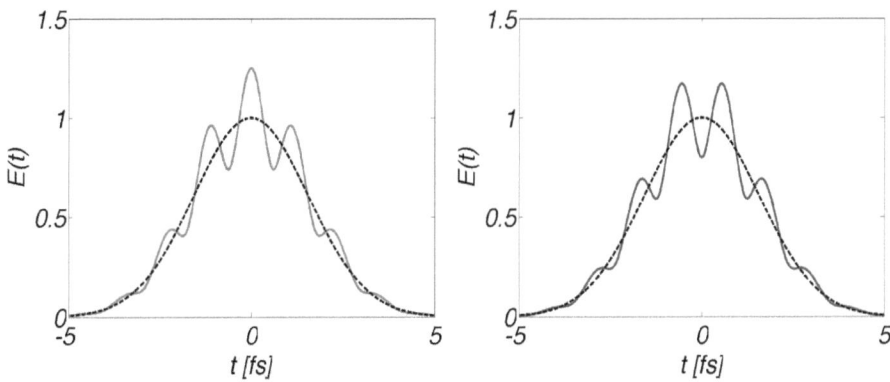

Figure 3-6: CEP dependence of the electric field depicted in Figure 3-5 in elliptically polarized light. The black broken line shows the temporal envelope of a circularly polarized pulse. On the left (red solid line), the temporal field evolution is shown for the envelope pointing into the direction of the major axis of the ellipse, corresponding to a CEP=0, on the right (blue, solid line), the envelope points along the minor axis of the ellipse, CEP= $\pi/2$.

3.3.4 Absolute value of the CEP in circular and elliptical light

In chapter 3.3.2 it was shown that the CEP in perfectly circular light merely rotates the pulse field around the propagation axis. A shift in CEP $\Delta\phi_{CEO}$ results into an equal shift in the pointing angle of the field maximum $\Delta\phi_{max}$, the value of the CEP can thus be directly identified with the pointing of the maximum of the electric field.

The definition of the absolute value of the CEP on the other hand is somewhat arbitrary in circular polarization. One possible definition is to choose a fixed axis in space, for example the polarization axis of the linearly polarized light that was used to create the circular light.

If an ellipticity is introduced, the polarization axis breaks the rotational symmetry. A natural choice is then to connect the definition of the absolute CEP value to the polarization ellipse: Since in an experiment it is almost impossible to attain perfectly circular light, for this thesis the latter definition will be used, so that a CEP of 0 corresponds to the case where the pulse's envelope maximum is aligned with the major axis of the polarization ellipse. The field shown in Figure 3-6 on the left then corresponds to a CEP=0, the field on the right to CEP=$\pi/2$.

Unfortunately in elliptically polarized light it is not straightforward any more to connect the angles of the ellipticity peaks to the CEP. Because the pointing of the ellipticity peaks is determined by both, the orientation of the polarization ellipse and the orientation of the pulse envelope, the final angle depends on the precise amount of ellipticity and envelope gradient.

This can be best seen for a CEP=$\pi/2$, i.e. where the envelope maximum points along the minor axis of the ellipse and the pulse ield shows two equal maxima. The pointing of the electric field is shown for different ellipticities and a field envelope with FWHM =3.7 fs in Figure 3-7.

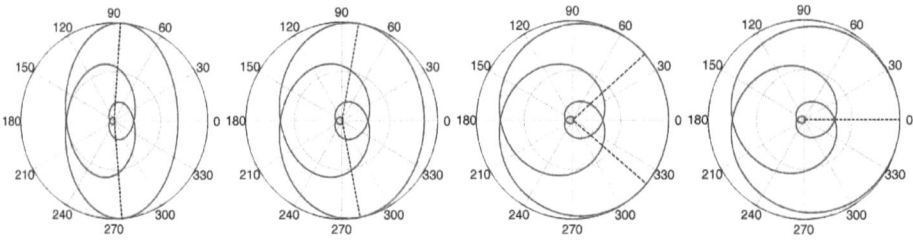

Figure 3-7 Shown is the evolution of an elliptically polarized pulse with a field envelope wit FWHM =3.7 fs in the polarization plane for different ellipticities. In blue, solid line the pulse's electric field is depicted. The polarization ellipse is oriented vertically (90 degrees), the electric field envelope points along the minor axis of the ellipse. Black dotted lines indicate the pointing of the electric field maxima. The ellipticities from left to right are: ε =0.8, 0.9, 0.98 and 1, i.e. circular, respectively.

3.4 Streaking

On a real clock-face, the ticks that map angle to time are equally spaced. In the atto-clock, the mapping function depends on the precise pulse shape. The most important condition for the mapping function is that it is strictly monotonic, so that particles that are ionized at different times always end with a different final angle.

The final momentum of an electron (or ion) in the electric field of a laser pulse can be calculated by integrating over the force exerted on the particle between the instant of ionization t_i, and infinity (when the pulse field has ceased). The momentum transferred to the particle by the x-component of the electric field $E_x(t) = E_{0x}(t) \cdot \cos(\omega_0 t)$ reads as:

$$p_x = \int_{t_i}^{\infty} q \cdot E_x(t) dt = \left[\frac{q}{\omega_0} E_{0x}(t) \sin(\omega_0 t) \right]_{t_i}^{\infty} - \frac{q}{\omega_0} \int_{t_i}^{\infty} \frac{dE_0(t)}{dt} \sin(\omega_0 t) dt \qquad (0.1)$$

The analysis of the mapping field is restricted to particles with no initial velocity, in particular no initial angular momentum.

If the second term in this expression is neglected (assuming a slowly varying pulse envelope and thus $dE_0(t)/dt \approx 0$), the sine dependence in the momentum shows a 90-degree phase shift compared to the cosine dependent electric field of the pulse. The same phase shift is obtained from the y-component of the pulse field, so that the final momentum of the particle points ahead of the electric field at the instant of ionization by 90 degrees and within one cycle the angle increases monotonically with ionization time.

For circularly polarized light the momentum gain can be calculated independently for the two perpendicular components of the field introduced in chapter 3.3.2

For few cycle pulses, the variation of the pulse envelope cannot be neglected. The mapping function is still monotonic if the second term in eq 1.1 is smaller than the first term for all ionization times.

In Figure 3-8, the streaking angle is shown as a function of the time of ionization in the pulse. For a ~ 5 3fs Gaussian pulse and perfectly circular polarization, the streaking angle rises linearly from around 87 degrees to 93 degrees in the two femtoseconds around the pulse center where ionization is significant (blue solid line). In the case of a small ellipticity (green dotted line) the streaking angle shows the same linear trend but an additional oscillation that leads to a variation in streaking angle of almost twelve degrees in the central two femtoseconds of the pulse.

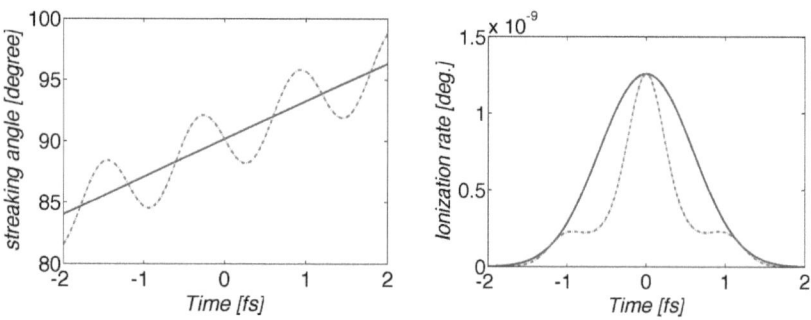

Figure 3-8 Left: Streaking angle for a Gaussian pulse for perfectly circular light (blue solid line) and an ellipticity of 0.048 (green dotted). On the right: corresponding ionization rates.

In this relation between streaking angle and time the CEP sets the angle of the temporal origin, so the CEP must be fixed to know the absolute mapping from time to angle.

The accuracy and resolution that can be achieved with AAS depend strongly on the specific implementation of the technique, e.g. if the ionization time is required for individual particles or if the streaking of an entire distribution is measured.

If the streaking of individual particles is considered, the full mapping function is required to determine the instant of ionization from the final momentum. In particular for this absolute streaking angle the CEP needs to be fixed (thereby fixing the position on the time axis in Figure 3-8). In reality however, e.g. in the case of a tunneling electron that appears with essentially zero kinetic energy in the continuum, a more severe constraint for the determination of the temporal origin of a single particle is posed by the uncertainty in a quantum mechanical measurement: The width of the electron wave packet was calculated to be around 30 degrees (see chapter 7.3), much more than the variation in streaking angle.

In the experiments considered in this thesis, it was not necessary to determine the ionization time of single particles independently. Instead the timings of the ionization rate distributions were measured showing features distinctly larger than the limit of temporal resolution (i.e. the ellipticity peaks). In this case the initial spread of the electron wave packet can be treated as a statistical error and averaging over many ionization and streaking events increases the temporal accuracy.

3.5 Comparison to energy streaking

Attosecond resolution was achieved for the first time with a similar technique to AAS, referred to as energy streaking [9, 48, 54]. In energy streaking linearly polarized light is used to map time to energy and it was used in combination with an attosecond trigger pulse.

In the experiment 'atomic transient recorder' [55], a 5fs long IR pulse was used to measure the duration of an attosecond pulse. At the same time the temporal evolution of the electric field of the IR pulse was mapped with an accuracy of 200 as. The principle is shown in Figure 3-9.

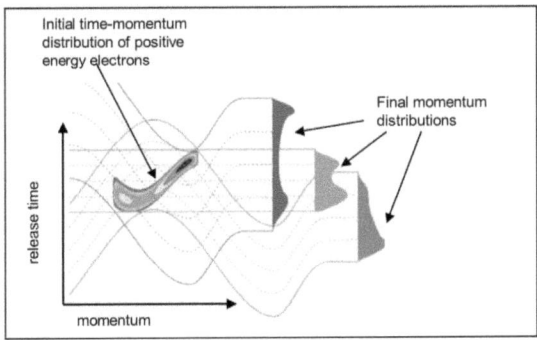

Figure 3-9 Adapted from [55]. The initial electron distribution is mapped depending on the timing between the attosecond pulse and the streaking field.

Atoms were first excited by an attosecond pulse. The initial time–momentum distribution of electrons is then created through ionization, e.g. in an Auger decay or through tunneling ionization. The electron distribution carries direct time-domain information about the excitation and relaxation dynamics of the electronic shell.

The electric field of the laser pulse that provides the streaking field is parallel to the direction of observation of the ejected electrons. At different delays different projections of the entire time–momentum distribution of the atomic electron emission are created in momentum space. These final momentum distributions are measured after the laser pulse left the interaction region.

Since in this case both the streaking field and the initial momentum distribution were unknown, different projections were measured by choosing different delays between the exiting pulse and the streaking field.

In the absence of the laser field, the electron momenta do not change after detachment. In the presence of a probing field, accumulation of electrons with a given final momentum occurs along the lines of constant canonical momentum (blue and red lines). The corresponding projections (streaked spectra) are represented in red and blue. From a suitable set of such tomographic projections the time–momentum distribution of electron emission can be retrieved, providing direct time-domain insight into atomic dynamics triggered by the XUV pulse synchronized to the probing laser field. Other than angular streaking that provides unique mapping over an entire cycle of the laser field, energy streaking is only unique within a quarter cycle of the electric field, the trigger event thus needs to be relatively short or several measurements at different delays need to be taken. Another difference between angular and energy streaking is the variation of the streaking field: While in circular light the streaking angle varies slowly with the envelope of the streaking pulse, in energy streaking maximum resolution is achieved for the highest field gradient, i.e. the zero crossings of the streaking field, while resolution is lowest at the peaks of the electric field oscillation, where field ionization by the laser pulse has the highest probability. Energy streaking is therefore not suitable to study tunneling ionization. On the other hand angular streaking depends on a narrow initial angular distribution.

Chapter 4

Semi-classical simulations

4.1 Introduction

In Chapter 3 it was shown that the momentum distributions forming the atto-clock can only be correctly interpreted by comparing them to simulations based on a precise model of the laser pulse:

The electrical field of the pulse determines the momentum spectra through the ionization rate and through the subsequent streaking. In a first approximation AAS can be treated as a 2-step process where the ionization step is treated separately from the subsequent acceleration in the field. Once the particle is set free, the Coulomb interaction between electron and remaining ion is neglected.

4.2 Pulse simulation

To simulate the temporal pulse field it is convenient to first treat spectrum and spectral phase separately in the spectral domain and then apply a Fourier transformation to convert to temporal domain. The spectrum and spectral phase of a linearly polarized pulse can be measured with SPIDER (see chapter 5.3). In general, the pulse is not measured at the location of the experiment, so that propagation through optical components and air needs to be accounted for by adding appropriate phase terms to the measured spectral phase. Providing the propagations lengths are known, the phase contributions of the main materials can be precisely calculated using Sellmeier equations. To simulate the propagation through the quarter wave plate the pulse field is split into two perpendicular components along the fast and slow axis to add the respective phase terms.

4.2.1 Phase and spectrum

For propagation through media it is most convenient to use the complex representation of the pulse's electric field in the spectral domain $\tilde{E}(\omega)$. The field can be split in the spectral amplitude $|\tilde{E}(\omega)|$ and the phase term with the spectral phase $\varphi(\omega)$:

$$\tilde{E}(\omega) = |\tilde{E}(\omega)| e^{i\varphi(\omega)}$$

The square of the spectral amplitude $I(\omega) \propto |E(\omega)|^2$ corresponds to the spectral power density, commonly referred to as spectral intensity or simply spectrum.

The frequency representation is connected to the time domain representation by a complex Fourier transform:

$$\tilde{E}(\omega) = \frac{1}{\sqrt{2\pi}} \int_{-\infty}^{+\infty} \tilde{E}(t) e^{i\omega t} \, dt$$

and the inverse transform back to the time domain:

$$\tilde{E}(t) = \frac{1}{\sqrt{2\pi}} \int_{0}^{-\infty} \tilde{E}(\omega) e^{-i\omega t} \, dt$$

If a pulse travels through a dispersive medium, i.e. a medium in which the refractive index n depends on frequency ω, an additional spectral phase $\varphi(\omega)$ is imposed on the pulse:

$$\varphi(\omega) = \frac{\omega}{c} n(\omega) L$$

L denotes the length of the dispersive medium and c is the speed of light in vacuum. Especially for very short pulses, which have a large bandwidth, the pulse is significantly distorted by dispersion.

The frequency dependent refractive index can be calculated with Sellmeier equations.

For quartzglass and air the Sellmeier equation reads:

$$n(\lambda[\mu m]) = \sqrt{1 + \frac{a_1 \lambda}{(\lambda - b_1^2)} + \frac{a_2 \lambda}{(\lambda - b_2^2)} + \frac{a_2 \lambda}{(\lambda - b_2^2)}}$$

with coefficients for glass:

$a_1 = 0.6961663 \quad b_1 = 0.0684043$
$a_2 = 0.4079426 \quad b_2 = 0.1162414$
$a_3 = 0.8974794 \quad b_3 = 9.896161$

and for air

$a_1 = 0.0048673 \quad\quad b_1 = 0.0204978$
$a_2 = 0.0000585831 \quad b_2 = 0.1320303$
$a_3 = 0$

Figure 4-1 shows the refractive index for quartz glass.

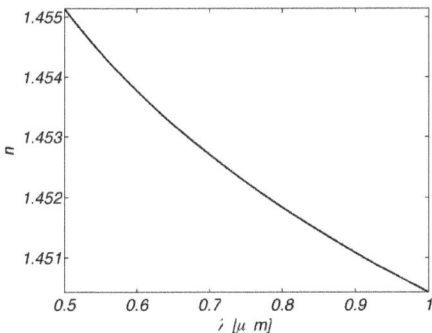

Figure 4-1 The refractive index of quartzglass in the wavelengt range around 800 nm

4.2.2 Quarter wave plate

The phase contribution of the quarter wave plate is particularly critical. The pulse is split into two pulses co-propagating inside the material. One pulse is slightly delayed compared to the other, to achieve perfectly circular polarization, the accumulated phase of the two pulses would need to differ by exactly $\pi/2$ as shown in chapter 3.3.2 for every frequency component in the pulse's spectrum.

The achromatic quarter wave plate consists of two birefringent plates made of 503.22 μm of MgF_2 and 637.23 μm of crystalline quartz that are mounted so that the fast axis of one plate is aligned with the slow axis of the other plate. The respective thicknesses are chosen to yield an overall phase delay around $\pi/2$.

To calculate the phase acquired by both parts of the pulse, again Sellmeier equations were used for both materials, however to calculate the phase difference that finally determines the degree of residual ellipticity, the Sellmeier equations are not precise enough, instead tables provided by the manufacturer were used.

For MgF_2, the Sellmeier equation from chapter 4.2.1 can be used with the following coefficients for the extraordinary ray [56]:

$$a_1 = 0.413440230 \quad b_1 = 0.0368$$
$$a_2 = 0.504974990 \quad b_2 = 0.0908$$
$$a_3 = 2.49048620 \quad b_3 = 23.7720$$

For the quartz plate a modified Sellmeier equation applies:

$$n = \sqrt{a_0 + a_1 \lambda + \frac{a_2}{\lambda} + \frac{a_3}{\lambda^2} + \frac{a_4}{\lambda^3}}$$

Here the coefficients for propagation along the ordinary axis were used:

$$a_0 = 2.3849$$
$$a_1 = -0.01259$$
$$a_2 = 0.01079$$
$$a_3 = 0.00016518$$
$$a_4 = -0.00000194741$$

The optical activity of quartz can be neglected for waves propagating perpendicular to the crystal axis [57]. The polarization eigenmodes, i.e. modes that do not change their polarization during propagation are slightly elliptical and the direction and length of the axes of the polarization modes are wavelength dependent.

To calculate effect of the $\lambda/4$ plate polarizations the Jones formalism is used. The polarization mode of the electric field is described as a Jones vector V_J:

$$E = \begin{pmatrix} E_x e^{-i(\omega t - \phi x)} \\ E_y e^{-i(\omega t - \phi x)} \end{pmatrix} = V_J e^{-i\omega t}$$

All optical elements can then be expressed as Jones-matrices acting on V_J.

The incoming polarization is projected onto the elliptical polarization eigenmodes of the wave plate. Then the phase shift can be applied and the polarization is projected back onto linear eigenmodes E_x and E_y.

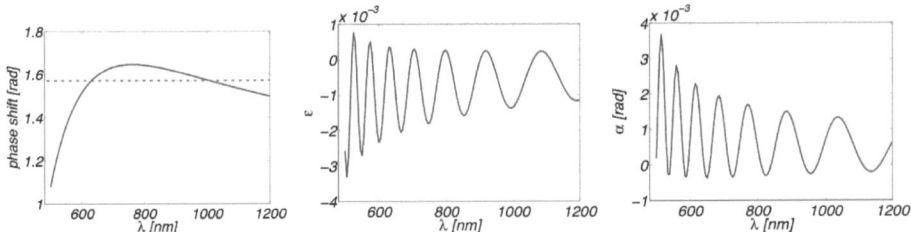

Figure 4-2 Data of the $\lambda/4$ plate. Phase, angle of the principal axis and ellipticity as a function of wavelength. The dotted line in the first panel indicates the $\pi/4$ phase shift required for perfectly circularly polarized light.

Figure 4-2 shows the parameters provided by the manufacturer that are needed to characterize the $\lambda/4$ plate. The first panel shows the phase shift depending on wavelength (solid line) and for comparison the ideal $\pi/2$ phase shift (dashed line). The second and third panels show the parameters ε and α, describing the ellipticity and angle of the polarization eigenstate of the wave plate as a function of wavelength. The elliptical eigenmodes of the quarter wave plate P_1 and P_2 then read:

$$P_1 = \begin{pmatrix} \cos(\alpha) & -\sin(\alpha) \\ \sin(\alpha) & \cos(\alpha) \end{pmatrix} \begin{pmatrix} \frac{1}{\sqrt{1+\varepsilon^2}} \\ \frac{i\varepsilon}{\sqrt{1+\varepsilon^2}} \end{pmatrix}$$

$$P_2 = \begin{pmatrix} -\sin(\alpha) & -\cos(\alpha) \\ \cos(\alpha) & -\sin(\alpha) \end{pmatrix} \begin{pmatrix} \frac{1}{\sqrt{1+\varepsilon^2}} \\ \frac{-i\varepsilon}{\sqrt{1+\varepsilon^2}} \end{pmatrix}$$

4.3 Streaking in the field

The streaking was calculated according to the formula in chapter 3.4, i.e. neglecting the influence of the Coulomb force between electron and ion during the propagation phase. Once the pulse field is known, streaking is easily calculated summing the force for incremental steps.

Under the assumption that the particle starts with zero velocity, the final momentum of an electron (or ion) in the electric field $\vec{p}(t_{0,ion})$ was calculated for each instant of ionization $t_{0,ion}$:

$$\vec{p}(t_{0,ion}) = q\left[\sum_{t=t_{0,ion}}^{\infty} E_x(t)\Delta t + \sum_{t=t_{0,ion}}^{\infty} E_y(t)\Delta t\right]$$

where q is the electric charge and $E_x(t)$ and $E_x(t)$ are the components of the electric field $\vec{E}_{ell}(t)$ of the Laser.

A simulation of AAS solving the time dependent Schrödinger equation (TDSE) performed by Harm Muller [58] was used to determine the effect of the Coulomb potential on the streaking angle. The attractive force of the Coulomb potential acts mainly in radial direction and leads to an intensity independent angular offset of 5-9 degrees depending on the ellipticity.

4.4 Ionization, ADK rates

In this work ionization was modeled using ADK [59] rates mainly because ADK theory provides analytical rate expressions in the regime of $\gamma < 1$. In principle ADK rates are cycle-averaged rates, but in this case were used as instantaneous rates and for $\gamma \sim 1$.

In the case of circularly polarized light the ADK rate assumes the following form:

$$w_{ADK}(t) = \frac{E(t)D(t)^2}{8\pi Z}\exp\left(-\frac{2Z^2}{3n^{*3}E(t)}\right)$$

$$D(t) \equiv \left(\frac{4eZ^3}{E(t)n^{*4}}\right)^{n^*}$$

where e is the Euler number, $E(t)$ is again the electric field of the pulse, Z is the charge after ionization and n* is the effective principal quantum number:

$$n^* \equiv \frac{Z}{\sqrt{2I_p}}$$

Figure 4-3 shows the ionization rate $w_{ADK}(t_{0,ion})$ for helium as a funcion of intensity in circularly polarized light.

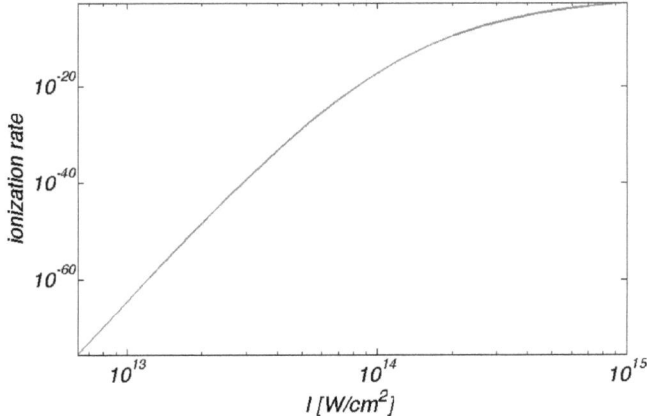

Figure 4-3 Ionization rates for circularly polarized pulses calculated with ADK

4.5 Intensity calibration

In the experiments presented in this thesis, the intensity was around $3 \cdot 10^{14} W/cm^2$ and thus far below saturation.

The calculated pulse field needs to be calibrated with the maximum intensity to yield the correct ionization rate distribution. Figure 4-4 shows the ionization rate in helium of the same elliptical pulse field of a ~5fs pulse for three different intensities: Since the ionization yield over the entire pulse is far away from saturation, the maximum intensity affects only the contrast.

The maximum intensity in the pulse can be calculated from the power in the beam if time duration and focal parameters of the beam are known. It is much more precise however to calibrate the final intensity directly through the momentum that the ions gain during the streaking in the field. This relation can be expressed for circulary polarized light in the following useful formula:

$$I_0 \left[10^{14} W/cm^2 \right] = 2.2776 \cdot p^2 \left[a.u. \right]$$

usually in the experiment, the momentum is naturally calculated in atomic units while the rate formula in the simulation are given in SI units.

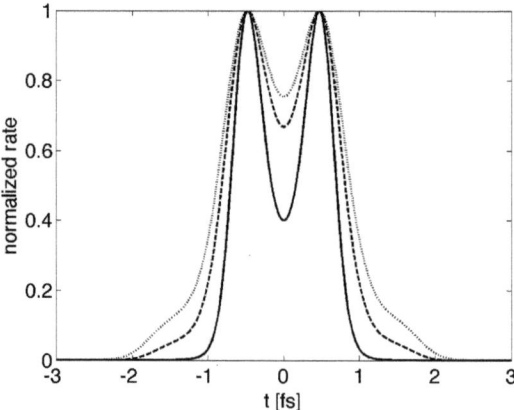

Figure 4-4 Ionization rate as a function of time for the same pulse field but different intensities of an elliptically polarized pulse calculated with ADK formulas. The rates are normalized to peak rates. The dotted line corresponds to a peak intensity of $1 \cdot 10^{15} \, W/cm^2$, the dashed line to $5 \cdot 10^{14} \, W/cm^2$ and the solid line to $1 \cdot 10^{14} \, W/cm^2$.

Chapter 5

Experimental setup I: The Laser Pulse

For both AAS experiments presented in this thesis (Chapter 7 and Chapter 8), elliptically polarized 5.9 fs pulses were used and helium ion momentum distributions were measured with a cold target recoil ion momentum spectrometer (COLTRIMS) [60]. In addition, CEP stabilization was required for the first experiment characterizing the accuracy of AAS.

The experimental setup can be split into two parts: The generation, characterization and manipulation of the laser pulses are treated in this chapter, the COLTRIMS apparatus for the ionization measurements is discussed in the next chapter.

The high intensity 5.5 fs pulses that were required for the experiments in this thesis cannot be generated directly from a laser amplifier. Instead ~30 fs pulses are produced by a CPA (chirped pulse amplification) system and then spectrally broadened in a filament compression setup, yielding a linearly polarized, essentially transform limited 5.5 fs pulse.

5.1 Laser

A short pulse in time is created through the coherent superposition of a large band of frequencies. To produce such a pulse a laser medium is required that allows gain over a large bandwidth. The most suitable solid-state medium for generating femtosecond pulses is titanium-doped sapphire (Ti:Sapphire) [61], with a gain bandwidth of about 230 nm at a central wavelength of around 800 nm.

For geometrical reasons a laser cavity allows only certain frequencies to propagate within the range supported by the gain medium. If these modes oscillate with random relative phase, the laser oscillator generates continuous light. Only if the relative phase of the different frequency components is fixed, i.e. phase-locked or mode-locked, at certain times all frequency components will interfere constructively, creating a pulse. This constructive interference occurs periodically forming a pulse train.

There are several techniques to achieve mode-locking, most commonly used for ultrashort Ti:Sapphire lasers is Kerr lens mode-locking (KLM) [62]. It employs the optical Kerr effect to establish a phase relation between adjacent laser modes without the need for external modulation (passive mode-locking). The Kerr effect depends nonlinearly on the intensity, so that the continuous, low intensity light is not affected. For short and therefore intense pulses in the cavity it creates a Kerr-lens, inducing self-focusing in the crystal. An aperture can then be introduced letting only the short pulses pass, so that eventually the continuous light will cease.

With Kerr lens mode-locked Ti:Sapphire lasers, sub-10-fs pulses can routinely be generated and even sub-6-fs pulses were produced directly from an oscillator [63, 64]. The typical repetition rate of Ti:Sapphire oscillators is 70-80 MHz and pulse energies are in the nJ range.

The commercial laser system[1] used for all experiments in this thesis is based on the CPA technique [65], that allows to boost the energy of the weak mode-locked oscillator pulses while preserving the CEP and the large bandwidth up to a spectrum supporting ~25 fs and the CEP. The principle is shown in Figure 5-1.

The pulse generated in the oscillator is sent several times through an inverted gain medium, again a Ti:Sapphire crystal, coherently adding photons to the pulse with each pass. The issue with short pulses is that peak intensities quickly reach the damage threshold of the gain material (i.e. 10 GW/cm²). The solution to this problem was to stretch the pulse in time before amplification while preserving its spectral bandwidth e.g. by sending it through a dispersive material.

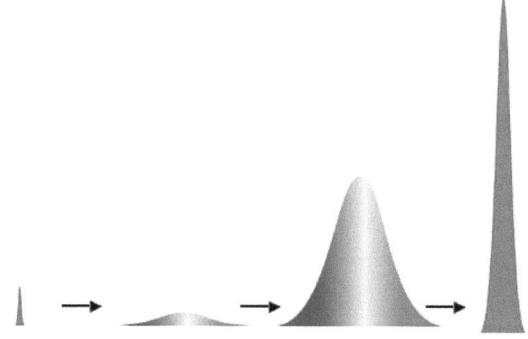

Figure 5-1 The essentially transform limited oscillator pulse is stretched by chirping it, then amplified and recompressed.

---

[1] Femtopower Compact Pro, Femtolasers GmbH

In the regime of normal dispersion (all glasses show normal dispersion in the optical wavelength region) the material adds positive dispersion to the spectral phase of the pulse, i.e. the high frequency components are delayed compared to the low frequency components in the pulse. The spectral components that form the pulse are now separated in time without loosing their coherence. After the amplification, the stretching process can be reversed by separating the frequency components spatially and tune their geometrical paths length to overlap the frequency components again in time. This can be done with e.g. with a grating setup, in this case a prism setup was used for its low energy losses since all prisms can be used under Brewster angle.

To achieve a transform-limited pulse, i.e. a pulse with perfectly flat spectral phase, the phase acquired in the stretching process must be matched with the phase from the compression setup. This matching can be greatly improved by using an acousto-optic programmable dispersive filter[2] that allows adding arbitrary phase terms to the pulse.

The laser setup consisting of oscillator, amplifier and compressor is shown in

Figure 5-2:

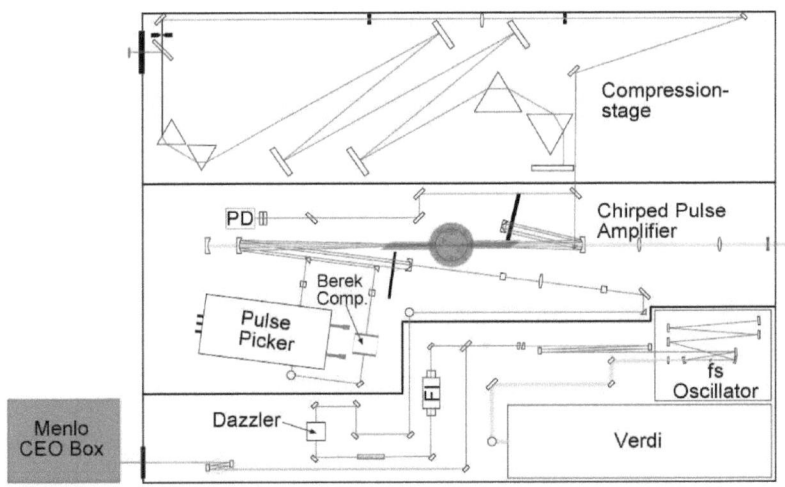

○ Periscope with Polarization Rotation
□ Periscope without Polarization Rotation

---

[2] better known by its brand name Dazzler, Fastlite

Figure 5-2 The laser system consisting of the fs-oscillator, amplifier and prism compressor. Pump light is shown as green, thick lines, the IR pulse is shown in red, thin lines. Only four passes through the amplifier crystal are visible since the pulse travel out of the plane.

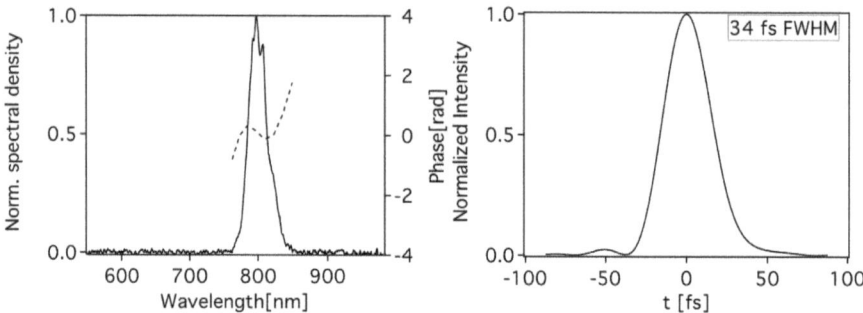

Figure 5-3 Spectrum and phase of a 34 fs pulse (left) and the temporal intensity envelope (right).

The pulses from the oscillator are sent four times through the amplifier crystal at the full repetition rate. The repetition rate is then strongly reduced to 1 kHz by pulse picking with a Pockels cell, that is triggered to let one pulse pass only every 1 ms. The pulse is then sent through the amplifier another 5 times reaching a final energy of around 1 mJ. After recompression in the prism compressor the pulse is transform limited with a pulse duration of 25-30 fs. Spectrum and phase of such a pulse are shown in Figure 5-3.

5.2 Pulse compression by filamentation

The pulse length that can be generated from the laser and amplifier is limited by the gain bandwidth of the medium to around 30 fs. The spectrum needs to be additionally broadened to support a 5.5 fs pulse. At the same time the temporal coherence of the frequency components needs to be preserved. This spectral broadening is achieved through filamentation [66] in a noble gas cell.

The most widely used technique for pulse compression is hollow fiber compression. The pulses are sent through a long guiding structure filled with a noble gas. Spectral broadening is achieved through self-phase modulation (SPM). This also induced a large linear chirp that is subsequently compensated by chirped mirrors. With this technique pulses as short as 4.5 fs [67] and even 3.8 fs [68]with a two stage setup were generated.

Due to the thin core alignment of hollow fibers is difficult and beam pointing instabilities directly translate to intensity and pulse width fluctuations.

Using filamentation removes these disadvantages. Other than using external guiding as in the hollow fiber, it relies on self-guiding in the gas. Two nonlinear (i.e. intensity dependent) effects are needed for the self-guiding: Self-focusing due to intensity-dependence of the refractive index of the medium, i.e. the Kerr-effect, and defocusing due to the formation of a free electrons [69]. The two effects are shown in Figure 5-4. These two effects balance each other and form a dynamic equilibrium leading to the formation of self-guiding filament with a core diameter of ~50 μm. Spatio-temporal reshaping of the pulse due to several highly nonlinear processes such as SPM (self-phase modulation) and plasma blue-shifting lead to a broadening of the spectrum.

The setup used is shown in Figure 5-5. The pulses are sent through two tubes filled with argon at a pressure of around 650 -900 mbar.

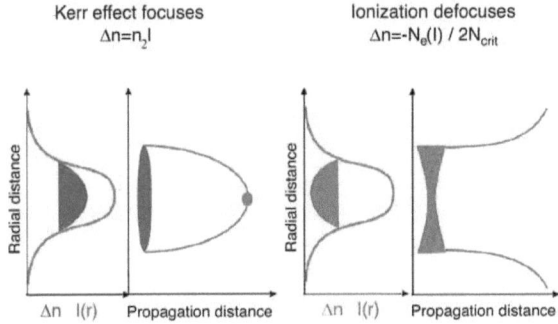

Figure 5-4 The two mechanisms that form a dynamic equilibrium leading to filamentation

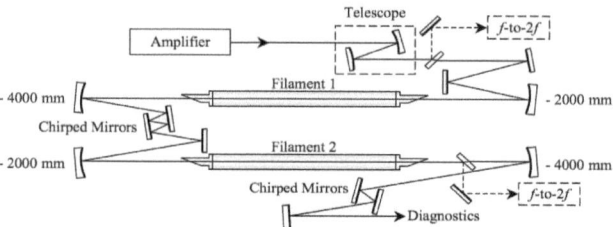

Figure 5-5 The two stage filamentation setup. The gas cells were filled with 650-900 mbar of argon. Chirped mirrors were used for compression.

In the first cell the pulse spectrum is broadened to support a 10 fs pulse. After recollimation and recompression with chirped mirrors it is focused into the second argon cell to broaden the spectrum even further to an ~5 fs pulse. Since the spectrum spans over an octave it can be directly used for a CEP measurement. The pulse is then pre-compensated for all material in the beam path by six bounces on chirped mirrors to yield an essentially transform limited pulse in the experimental region. Measurements indicate that also self-compression is achieved at least in the second filament.

During filamentation, the center wavelength is shifted to 725 nm and the spectrum is deformed. The filaments act also as mode cleaners.

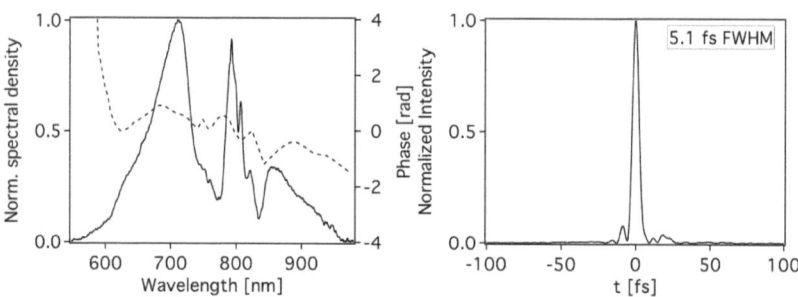

Figure 5-6 Left: Spectrum (solid line) and phase (broken line) of a 5 fs pulse. Right: The temporal intensity envelope

5.3 Pulse characterization: Spider

It is not possible to resolve the temporal pulse field directly in the time domain. Instead the spectrum of the pulse and the phase between the spectral components can be measured separately. This data can then be Fourier transformed to the temporal domain to yield the electric field of the pulse.

From this measurement the pulse at the point of interaction can be calculated by adding the dispersion for all material in the beam path between the characterization setup and the COLTRIMS using the formula introduced in 4.2.1 and 4.2.2. For fine adjustment of the dispersion, insertable glass wedges were used.

Several techniques exist for the measurement of pulses in the femtosecond domain with countless variants. The two main techniques are FROG (frequency resolved optical gating) [70] and SPIDER (Spectral interferometry for direct E-field reconstruction) [71-73]. In this thesis SPIDER [74] was exclusively used for pulse characterization.

SPIDER is a self-referencing technique using spectral interferometry between the pulse and a frequency shifted copy that is delayed in time [73].

The interferogram $I_{SPIDER}(\omega)$ then contains the phase information depending on the spectral offset or spectral shear $\delta\omega$ between the copies:

$$I_{SPIDER}(\omega) = |E(\omega)|^2 + |E(\omega + \delta\omega)|^2 + 2|E(\omega)E(\omega + \delta\omega)|\cos[\varphi(\omega + \delta\omega) - \varphi(\omega) + \omega\tau],$$

where $E(\omega)$ is the electric field, $\varphi(\omega)$ is the spectral phase of the pulse and τ is the delay between the two replica.

From the cosine term, the phase information can be extracted by the following, purely algebraic, method: the Fourier transform to the time-domain of the SPIDER interferogram consists of a peak at zero time and two side-peaks located near τ and $-\tau$. These peaks contain equivalent phase information [73].

One of the peaks is isolated by applying a suitable filter function and an inverse Fourier transform back to the frequency domain yields a complex function $c(\omega)$, the complex phase of which gives access to the spectral phase of the pulse:

$$\arg(c(\omega)) = \varphi(\omega + \delta\omega) - \varphi(\omega) + \omega\tau.$$

The large linear phase term $\omega\tau$ is given by an experimental parameter, the delay between the two copies. This linear phase term can be obtained from a separate measurement by spectral interferometry of the short unsheared pulse replicas and subtracted from the term above:

$$\Delta\varphi(\omega) = \varphi(\omega + \delta\omega) - \varphi(\omega).$$

From an arbitrarily chosen starting frequency ω_0 the spectral phase $\varphi(\omega)$ is calculated at evenly spaced frequencies $\omega_i = \omega_0 + i \cdot \delta\omega$ by the following concatenation procedure:

$$\varphi(\omega_0) = \varphi_0$$
$$\varphi(\omega_1) = \Delta\varphi(\omega_0) + \varphi(\omega_0) = \varphi(\omega_0 + \delta\omega) - \varphi(\omega_0) + \varphi(\omega_0) = \varphi(\omega_0 + 1 \cdot \delta\omega)$$
$$\vdots$$
$$\varphi(\omega_{i+1}) = \Delta\varphi(\omega_i) + \varphi(\omega_i).$$

The constant φ_0 remains undetermined but is only an offset to the spectral phase that does not affect the temporal pulse shape, and it can thus be set equal to zero. The spectral phase can then be written as:

$$\varphi(\omega_{i+1}) = \sum_{k=0}^{i} \Delta\varphi(\omega_k).$$

The optical setup used for the SPIDER measurement is shown in Figure 5-7. At the surface of the glass block, the pulse is split into a reflected part, the 'short pulse', indicated as a dotted line, and the 'long pulse', indicated as a solid line that is transmitted through a highly dispersive glass block strongly stretching the pulse. This stretched pulse provides the spectral shear.

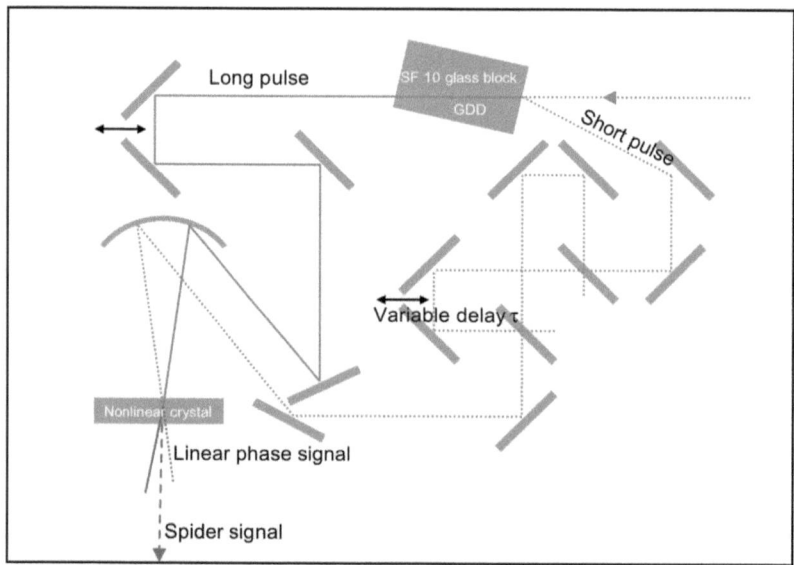

Figure 5-7 SPIDER setup: Short pulses are depicted as dotted lines, the stretched pulse is depicted as a solid line.

The reflected, short pulse is then sent into a Michelson interferometer where it is split into two copies. One copy is slightly delayed in the order of 200 fs before the pulses are recombined on a second beamsplitter to proceed collinearly.

The two short pulses are recombined with the stretched pulse in a nonlinear optical crystal under a small angle. This leads to three spatially separated signals around 400 nm, half the wavelength of the fundamental pulse at 800 nm. Trough nonlinear interaction in the crystal, the long pulse as well as the short pulses are doubled. The latter signal contains the information about the linear phase term $\omega\tau$.

The central signal emerging between the two frequency-doubled beams is the SPIDER signal containing the phase information as indicated by a broken line in Figure 5-7. It stems from the upconversion of the two short pulses by sum-frequency generation with the strongly stretched pulse.

The dispersion in the glass block leads to a time dependent nonlinear frequency in the stretched pulse, so that the instantaneous frequency that is used for upconversion depends on the relative time between the stretched and the short pulse. This means, that the delayed pulse is upconverted with a different frequency than the original short pulse.

The delay τ between the two short copies determines the positions of the two side-peaks of the Fourier transform of the interferogram. It is therefore chosen so that the side-peaks are well separated from the center-peak. On the other hand the fringe spacing of the interferogram is proportional to $2\pi/\tau$ and thus τ must be sufficiently small that the spectrometer is able to fully resolve the fringes. The stretching factor GDD, that is determined by the dispersion acquired in the glass block is then chosen such that the spectral shear $\delta\omega$, which determines the sampling interval of the reconstructed spectral phase, is small enough to ensure correct reconstruction of the electric field in the time domain according to the Whittaker-Shannon sampling theorem [75]. The variables τ, $\delta\omega$ and GDD are connected:

$$\delta\omega = \frac{\tau}{GDD}$$

This constrained relationship means that with a particular SPIDER setup, only pulses with a limited range of pulse durations can be measured.

An additional independent measurement of the pulse spectrum provides all the necessary information to determine the electric field in the time domain by a Fourier transform of:

$$E(\omega) = \sqrt{I(\omega)}e^{i\varphi(\omega)},$$

where $I(\omega)$ is the measured spectrum and $\varphi(\omega)$ is the reconstructed spectral phase. The electric field in the time domain is thus determined unambiguously except for the CEP.

5.4 CEP stabilization

As mentioned in chapter 3.3, without CEP stabilization the offset between the carrier phase and the envelope of the laser pulse fluctuates from pulse to pulse.

The reason for these fluctuations is changes in the dispersion inside the oscillator cavity, rapid fluctuations as well as long-term drifts. These changes are caused by environmental parameters such as temperature changes and airflow as well as laser parameters such as pump power fluctuations. It is however possible to control the dispersion in the oscillator by modulating the power of the laser that pumps the oscillator crystal: The pump laser beam is sent through an acousto-optic modulator (AOM), that diffracts a portion proportional to the power of the acoustic wave created in the AOM of the beam onto a beam dump. The remaining power in the 0th order of the beam thus can be modulated and is then used for pumping the oscillator crystal.

The CEP stabilization setup that delivers the control signal consists of two feedback loops based on the so-called f – 2f technique. It is based on interference between the short wavelength tail (with frequency f) of an over octave spanning spectrum and the frequency doubled part of the long wavelength tail (with frequency 2f). This technique allows no absolute measurement of the CEP, only relative changes of the CEP can be measured.

The two loops are based on slightly different principles: The first loop measures the CEP slippage from pulse to pulse in the pulse train directly from the oscillator at the repetition rate $f_{rep} = 80 MHz$.

In the frequency domain this pulse train corresponds to a frequency comb with modes spaced by f_{rep} as shown in Figure 5-8. The CEP (f_{CEO}) is the offset of this comb on the frequency axis so that the frequency v_m of any mode m in the comb can be expressed as

$$v_m = f_{CEO} + m \cdot f_{rep}.$$

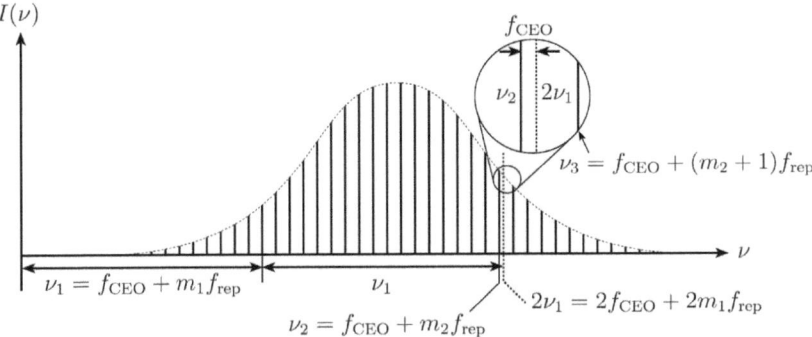

Figure 5-8 f-to-2f scheme for determining the CEO frequency of a pulse train. [3]

If this comb is frequency doubled a new comb is created with modes of the following form: $2\nu_m = 2f_{CEO} + 2m \cdot f_{rep}$. If the spectrum is sufficiently broad (i.e. more than octave spanning), there is spectral overlap between the fundamental comb and the frequency-doubled comb. The beating frequency between two neighboring modes of the two combs then yields f_{CEO} as shown in Figure 5-8.

The beating signal is measured with a commercially available setup[4]: the spectrum is first broadened in a nonlinear fiber to cover an octave and then split into two pulses by a dichroic beam splitter. The long wavelength part of this spectrum is frequency doubled and overlapped again with the short wavelength wing. The beating signal between the short wavelength wing of the original spectrum and the doubled long wavelength wing is measured with an avalanche photodiode. While this yields the CEP at the point where it is measured, it does not yield the absolute value of the CEP inside the laser oscillator, but with this signal, the pump power in the oscillator can now be modulated to keep the beating frequency fixed.

---

[3] Picture taken with permission from F. W. Helbing, Ph.D thesis, ETH 2004
[4] Menlosystems GmbH

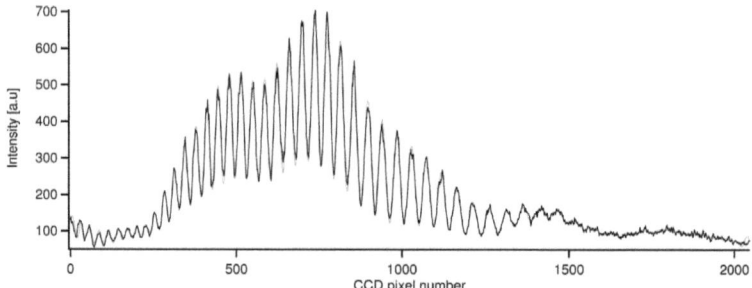

Figure 5-9 Measured (black trace) and reconstructed (gray trace) single-shot CEO spectral interference pattern. The reconstructed pattern was calculated using the phase($\psi\,(\omega)$) and amplitude information obtained from the Fourier filtering phase reconstruction technique. [5]

The second loop should account for CEP drifts during the amplification process and also the filamentation stages. Conveniently, the spectrum generated in the second filament is already octave spanning and thus needs no further broadening. After the amplifying stage the repetition rate of the laser is only 1kHz, so instead of a pulse train the pulses can be treated as isolated, consequently showing a continuous spectrum.

The CEP of a single pulse can be measured with spectral interferometry: For this loop a small portion of the beam is split off by reflection off a glass plate. Again the long wavelength tail is doubled in a nonlinear crystal and interfered with the short wavelength tail of the same spectrum, creating a spectrum that is modulated with the CEP. The spectral position of the interference pattern yields a signal that can be added as a slow feedback to the feedback signal of the fast loop in the modulation of pump power.

---

[5] Picture taken with permission from F. W. Helbing, Ph.D thesis, ETH 2004

Chapter 6

Experimental setup II: COLTRIMS

6.1 Introdution

COLTRIMS stands for COLd Target Recoil Ion Spectrometer [60] and the term describes the essential features: COLTRIMS measures the momentum of individual charged particles in three dimensions, yielding the full kinematic information available from a reaction.

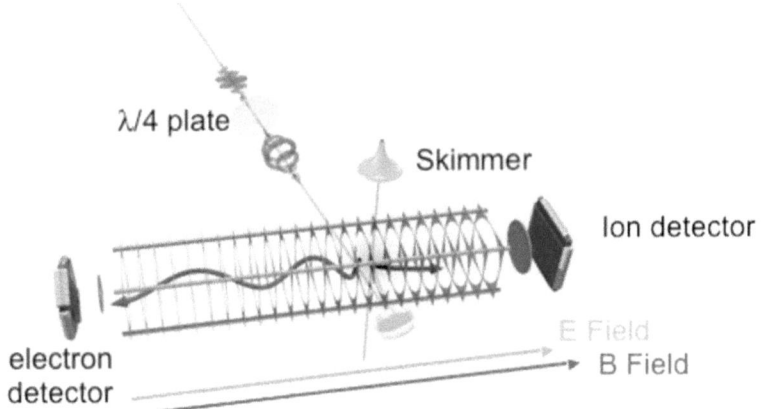

Figure 6-1 Coltrims setup showing the gas jet, laser and spectrometer with detectors for ions and electrons

Essential for high momentum resolution at low absolute momenta is the 'cold target', realized as a gas jet with very uniform momentum distribution. An outstanding feature COLTRIMS is also the possibility to measure several fragments (both electrons and ions) stemming from the same original atom or molecule in coincidence. This means that reaction channels with very low probability become accessible by filtering the data according to specific reaction products, energies or momentum vectors, thus offering a very high dynamic range.

The setup is sketched in Figure 6-1, the laser beam, the target gas jet and the 'spectrometer axis' set three mutually orthogonal axes.

6.2 Gas Jet

The gas jet is generated through super sonic expansion through a small nozzle (~20 μm) into a vacuum chamber at a background pressure of ~10^{-5} mbar. The gas is pre-cooled through contact with the nozzle that is cooled by liquid nitrogen to a temperature of ~-140 °C. In the supersonic expansion, a 'zone of silence' with very low internal temperature forms under the nozzle as shown in Figure 6-2 (taken from [76]).

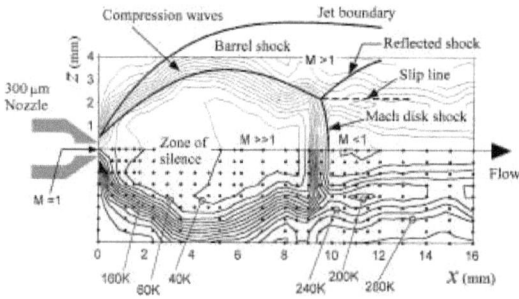

Figure 6-2 Raman mapping of rotational temperatures in a supersonic jet of CO2 under a stagnation pressure of 2 bars. Isothermal lines are depicted at steps of 20 K.

This zone of silence is cut out from the jet by passing through a skimmer with an orifice of 200 μm into a second vacuum chamber with lower background pressure. It then passes through an aperture into the main reaction chamber. This aperture consists of two adjustable razor blades and can be used to further collimate the gas jet, this second chamber also serves as differential pumping stage since the required background pressure in the main chamber is below $5 \cdot 10^{-10}$ mbar. After passing the interaction zone where the jet intersects with the laser, the gas jet is dumped directly into a turbo pump to keep the background pressure in the main chamber as low as possible.

Critical for the formation of a zone of silence and its dimension is the nozzle diameter d_{nozzle} and the ratio between the gas pressure P_0 of the nozzle and the background pressure P_b of the expansion chamber. The length of the zone of silence is given by

$$l_{zos} = \frac{2}{3}\sqrt{\frac{P_0}{P_b}} \cdot d_{nozzle}$$

The internal temperature of the gas can then be estimated by

$$T_{jet} = \frac{5}{2} \cdot \frac{T_{nozzle}}{S^2},$$

where S is the speedratio. It is given as an empirical function of the parameters T_0, P_0 and t_{nozzle}.

More information and a detailed overview over supersonic gas jets in general can be found in [77].

6.2.1 Spectrometer

The COLTRIMS apparatus has separate detectors for negatively and positively charged particles that are aligned on the spectrometer axis perpendicular to both the gas jet and the direction of the laser as shown Figure 6-1. Particles created in the interaction region at the intersection between the laser and the gas jet are drawn towards the detectors by a constant electric field along the spectrometer axis. Overlaid is also a magnetic field along the spectrometer axis, ensuring collection of all reaction products (up to a maximum initial momentum) from the full solid angle. The magnetic field is important only for the light electrons, forcing them on helical trajectories along the magnetic field while the much heavier ions only experience a small rotation.

Two large coils in Helmholtz geometry outside the vacuum chamber generate the magnetic field. On the axis trough the coil centers the magnetic field is given by:

$$B(z,n,I,r,d) = \frac{\mu_0 I n r^2}{2} \left(\left[\frac{1}{r^2 + \left(\frac{d}{2} - z\right)^2} \right]^{\frac{3}{2}} + \left[\frac{1}{r^2 + \left(\frac{d}{2} + z\right)^2} \right]^{\frac{3}{2}} \right),$$

where n is the number of windings of each coil, I the current through the coils, r their radius and d the distance.

The force of the electrical field is given by $\vec{F}_E = q \cdot \vec{E}$ while the Lorentz force exerted on the particles by the magnetic field is given by $\vec{F}_L = q \cdot \vec{v} \times \vec{B}$. If the two fields are aligned parallel, the two forces are orthogonal and can be treated seperately. To achieve this, the

Helmholtz coils need to be carefully aligned, in particular to also compensate for the earth magnetic field.

6.3 Detection

To reconstruct the momentum vector of a particle in all three dimensions, three independent coordinates need to be measured: in this case time of flight and position of impact on the detector plane. The detector consists of a large area MCP (micro channel plate) and a delay line detector [78]. The MCP serves a double purpose delivering the TOF signal and acting as an electron multiplier for the delay line detector. When a particle hits the MCP, it creates a small drop in the high voltage between the front end and the back end of the MCP by creating a cascade of electrons when passing through the MCP. This voltage drop can be measured and gives the time of impact on the detector. The emerging charge cloud travels on to the delay line detector that delivers the information about the point of impact.

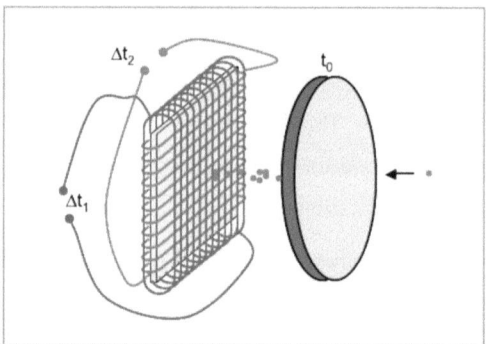

Figure 6-3 Detector with MCP (right) and delay line anode (left). A charged particle (dot on the right) hits the MCP generating an electron avalanche measurable as a small voltage dip that can be read and gives the time of impact. The charge cloud exiting the MCP hits the delay line anode at a specific point and creates electrical signals traveling along the wires. The measured travel time can then be transformed to a position information.

In the simplest case it consists of two perpendicularly oriented layers of wire pairs, the reference wire and the signal wire that is on a slightly positive potential of around 40 V compared to the reference. When the charge cloud hits the wire a voltage dip travels on each of the signal wires in both directions, the corresponding reference wires supply the ground. Comparing the arrival times on the ends of each wire then yields the position of impact.

The spacing between the wire pairs is 2 mm. For the electrons a more complicated geometry with three wire pairs in a hexagonal configuration is employed. In principle this provides redundant information, so the additional information can be used to reduce dead times and improve multi-hit capability.

6.4 Signal processing

The prerequisite for digitalization are as clean signals as possible from all components. In order to minimize reflections at connectors, careful impedance matching is crucial. The fast signals are decoupled from the high voltage over the detector by suitable capacitors. One issue in the digitalization is the high variation in signal amplitude that would lead to high timing jitters if a constant trigger level would be employed. Instead the pulses are fed into a constant fraction discriminator. There the pulse current is split up in two copies. One is amplified, inverted and delayed by a fixed amount. Both copies are then recombined again so that the resulting pulse has a zero-crossing at a constant fraction of the original pulse height. At this zero-crossing a logical NIM-(nuclear instrument methods)-pulse is triggered. This can then be fed into a TDC (time to digital) card in the computer that bins all events with a bin size of 0.5 ns. Once the data is digitized it can be further processed by special software [6].

6.5 Calibration

To be able to reconstruct the initial momenta and energies of the reaction's fragments without the effect of the detector fields E_D and B_D from the timing signals delivered by the detectors, careful calibration of E_D, B_D and t_0, the instant of ionization is essential.

The magnetic field affects the light electrons much more than the heavy ions so for the electron detection 'good' settings and calibration of the detector fields is less straightforward. The Lorentz force created by the magnetic field forces the electrons on helical trajectories. All electrons start at the instant of ionization and from the spatially

---

[6] Cobold by RoentDek Handels GmbH

confined focal spot of the laser but with different initial momentum vectors. This leads to a periodic refocusing of all electrons after completion of a full turn in the magnetic field as shown in Figure 6-4.

Note that the refocusing takes place in time, not along the z axis, the distribution of time of flight values according to the initial energy of the electrons is not affected by the magnetic field, since it is parallel to z. This periodic refocusing collapses the distribution in the x-y plane for certain initial energies to one pint on the detector, leading to very poor momentum resolution.

Depending on average momentum and the width of the momentum distribution in an experiment, both magnetic and electric field need to be adjusted to achieve maximum resolution (assuming a fixed position and size of the detectors).

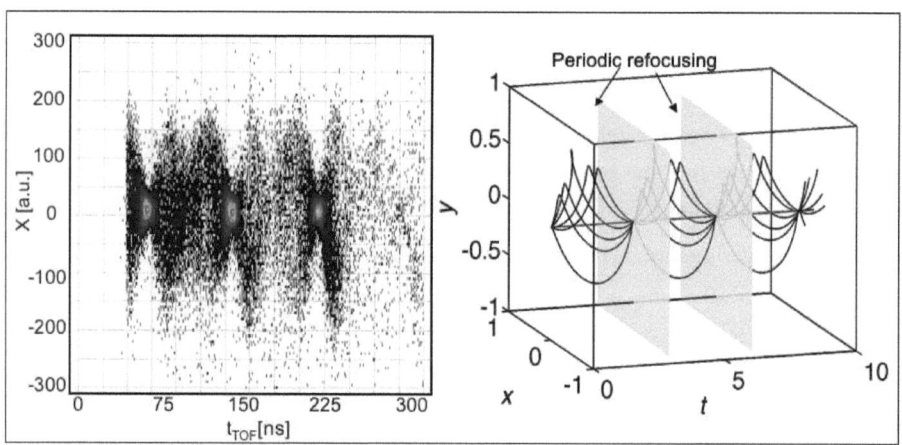

Figure 6-4 the left panel shows electron data. The distribution in the x-direction of the detector is plotted versus the time of flight of the electrons, clearly showing the refocusing in time caused by the magnetic field of the detector. On the right some electron trajectories are plotted, in light blue the refocusing times are marked.

The higher the magnetic field, the higher the maximum allowed momentum in the x-y plane that can still be imaged onto the detector. On the downside, a higher magnetic field means a smaller roundtrip time of the electrons with a higher number of points with no momentum resolution in the x-y plane. The electric field can be used to move the electron distribution between these focusing points, a higher electric field also narrows the time of flight of a given momentum distribution, leading to less resolution in the z coordinate.

The calibration steps are described in the order they were performed for this experiment. Before any calibration is performed, events are filtered according to the 'time sum' of the delay line signals. No matter where the particle hit this detector, the sum of the traveling signals should be constant. This ensures that only real events are taken into account.

The refocusing points can also be used to align the Helmholtz coils: If the earth magnetic field vector is perfectly compensated for, all temporal refocusings should arrive on the same spot on the detector and form a straight line in time. This can be visualized in a so-called wiggle plot as shown in Figure 5-6 where the one spatial coordinate of the detector is plotted vs. time of flight.

From the time difference Δt between two points of refocusing with time t_1 and t_2, the magnetic detector field B_D can be calculated with the following formula:

$$B_D = \frac{2\pi}{m_e e} \Delta t$$

From an educated guess about the number of turns the electrons have completed at t_1 the temporal origin of all electrons, t_0 can be calculated.

From the known distance between the focal spot and the electron detector d_e the electric field can be calculated using electrons that started with zero initial momentum in the z direction. (Such electrons can easily be created for example by ionization with light polarized along the y direction).

The electric field then reads:

$$E_D = \frac{2 d_e}{t_{TOF}^2} \frac{m_e}{e}$$

For the ions the temporal origin can be found by comparing the time of flight of different species. Usually apart from the target gas, hydrogen is present and also traces of oxygen and nitrogen.

Last also the spatial origin (x_0, y_0) of the detector needs to be set, it is easily found from the center of the electron or ion distributions. In particular for ions this needs to be set for each atomic species separately, thereby taking into account the momentum component along the y direction carried by the gas jet.

Once both fields and the temporal origin is known, momentum vectors can be reconstructed in all three dimensions by the following formulas:

The initial momentum in z direction is not altered by the magnetic field:

$$P_z = mv_{0,z} = m\left[\frac{s}{t_{TOF}} - E_D \frac{q}{2m} t_{TOF}\right]$$

$$P_x = \frac{-qB_D \cdot \{\sin(qt/mB_D) \cdot x + y \cdot [\cos(qt/mB_D) - 1]\}}{2 \cdot \cos(qt/mB_D) - 1}$$

$$P_y = \frac{-qB_D \cdot \{\sin(qt/mB_D) \cdot y - x \cdot [\cos(qt/mB_D) - 1]\}}{2 \cdot \cos(qt/mB_D) - 1}$$

From the momenta, the energy spectrum is easily calculated by

$$E = \sqrt{\left(\frac{p_x^2}{2m}\right)^2 + \left(\frac{p_y^2}{2m}\right)^2 + \left(\frac{p_z^2}{2m}\right)^2}$$

Chapter 7

CEP measurement with AAS

Measuring the CEP with the momenta of electrons created in strong field ionization was proposed in [79] for both linearly and circularly polarized light. Following this proposal the absolute value of the CEP was measured for the first time with circularly polarized light. In the experiment the spatial asymmetry of ionization was measured comparing energy spectra from two opposing detectors in a so called 'Stereo-ATI'-setup [80]. Since many atoms were ionized per laser pulse, the distribution could be measured in a single laser shot and no CEP stabilization was necessary. Later on CEP measurements were done in linearly polarized light where the CEP could be extracted from high energy electrons created through recollisions, a process that only takes place in linear light. These spectra are much more difficult to interpret, and the recollision process must be precisely calculated.

The original proposal to measure the CEP used perfectly circular light, where the peak of the ion momentum distribution can be directly linked to the CEP as described in chapter 3.3.2.

As explained in Chapter 3, the experimentally unavoidable ellipticity as well as the streaking in the pulse field complicates the relation between the CEP of the pulse and the resulting ionization distributions. In this chapter it will be shown that it is nevertheless possible to measure the CEP with very high accuracy with elliptically polarized light.

At the same time the measurement resolving the CEP can be used to test the accuracy of the AAS technique by comparing the experimental results to a simulation.

In the experiment presented in this chapter, the CEP was stabilized and then slowly ramped from pulse to pulse over almost 2π. From this slow ramp of the CEP, the angular position of the ellipticity peaks was extracted as a function of CEP. These angular positions were then compared to the to a semi classical simulation, yielding an accuracy of ~23 as. Experimentally the ellipticity peaks separated by half a cycle (around 1.2 fs) are well resolved and no faster features in time are expected either from the pulse field or from inherent ionization dynamics. The final limit to the resolution of AAS can thus not be derived from this data. Instead it is therefore deduced from estimates of the experimental resolutions and theoretical calculations. It turns out that the main limitation in this experiment is the inherent width of the electron wave packet corresponding to around 200 as.

7.1 Experimental setup

Linearly polarized 5.5-fs pulses were generated from 30-fs pulses in a two-stage filament compressor described in chapter 5.1. In addition to the stabilization of the CEP described in chapter 5.4, a slow ramp was added to the feedback signal, shifting the CEP shot by shot for a ramp over 2π over two hours. The CEP was measured after the filament compressor and recorded.

Figure 7-1 shows the in the upper panel measured CEP values over time in a levels of gray coded histogram. In the lower panel the corresponding error signal that was fed to the AOM controlling the CEP is shown over the time of the measurement. During the first 30 minutes almost no additional offset was given to the error signal showing that the inherent drift in the signal was already creating the desired ramp. After thirty minutes the system stabilized (visible also in the sharper distribution of the CEP value histogram) and the CEP ramp was driven by a ramp in the error signal.

The linearly polarized pulses before the quarter wave plate were fully characterized with a single-shot SPIDER set-up [81]. Close to circularly polarized light was obtained with the ultra broadband quarter wave plate described in chapter 4.2.2. The electric pulse field calculated from the measured SPIDER data was numerically propagated through all the material in the beam path yielding a period of 2.4 fs at the peak intensity and an approximate ellipticity of $E_x(t) / E_y(t) = 0.92$. This ellipticity value is only a simple estimate because for such ultra broadband pulses the ellipticity is not a constant parameter but changes from cycle to cycle in the pulse.

The momentum distributions of helium ions were measured with the COLTRIMS setup explained in Chapter 6. Per second around 10 helium ions were measured compared to ≈300 events/s from background gases. Due to their different mass other gasses are easily separable from the target ions.

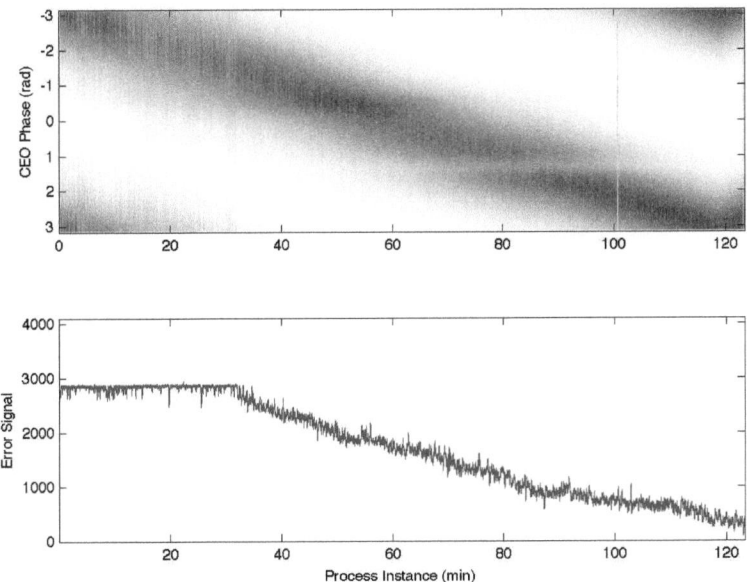

Figure 7-1 The upper panel shows the CEP values over measurement time for each single shot as a gray level encoded histogram, darker colors correspond to a higher density of measured CEP values. The center of the distribution is the CEP value, the width of the distribution is the uncertainty in CEP. The lower panel shows the corresponding error signal for the feedback, slowly shifting the phase shot by shot. The full range of the feedback signal is 4095 steps, in the first 30 minutes the error signal was almost constant meaning that the system was drifting.

7.2 Data

Figure 7-2 gives an overview over the data: Panel **a** shows helium ion momentum distributions projected onto the x-y polarization plane for four values of the CEP. The dataset containing the full ramp of the CEP was split into 36 subsets, so that each subset effectively presents a CEP ramp over 10 degrees.

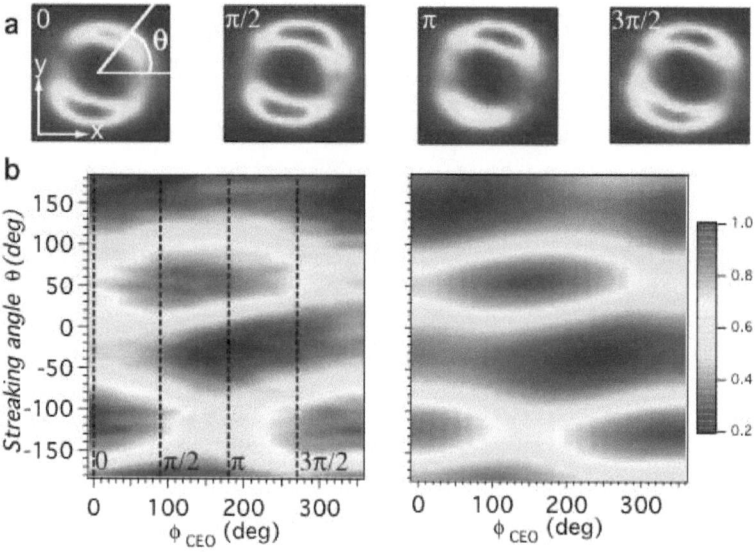

Figure 7-2 Overview over the measured helium ion momentum distributions while scanning the CEP over 2π and comparison with a semi classical simulation. The top row shows measured momentum distribution for four different values of the CEP. In the lower panel the momentum distributions are radially integrated and their angular dependence on the CEP is shown for the full scan of the CEP over 2π for both data (on the left) and simulation (on the right).

Clearly visible are the two peaks on the distribution stemming from the polarization ellipse. The first graph (CEP=0) corresponds to a situation where the pulse envelope points to the bottom left and is aligned with the polarization ellipse. Increasing the CEP rotates the envelope anticlockwise: In the second distribution, where the CEP was shifted by $\pi/2$ compared to the first graph, the pulse envelope was pointing along the minor axis of the ellipse (lower right corner in the distribution), creating an even distribution of ionization on both ellipticity peaks. Graph three and four, at CEP values of π and $3\pi/2$ are the mirror images of graphs one and two, respectively.

The full CEP ramp is shown in Figure 7-2 **b**. The momentum distributions in the polarization plane were radially integrated yielding angular momentum distributions for individual CEP steps. In this graph the CEP dependent intensity distribution between the two ellipticity peaks is clearly visible, as well as the slight CEP dependent angular shift of both distributions.

7.3 Resolution and accuracy

To investigate the temporal limitations of AAS, two different parameters need to be taken into account: Temporal resolution and accuracy. The resolution is given by the minimum time difference two features are allowed to have to be still separable. For the experimental temporal resolution different experimental errors have to be considered: The detector resolution, the temperature spread, fluctuations in the CEP value etc.

The accuracy is the error of the measurement compared to the real distribution, in this case the angular position of the ellipticity peaks is tracked in both the simulated and the measured distribution. It is limited both by statistical errors and possible systematic errors:

On the one hand the accuracy is limited by the statistical quality of the data: The longer the acquisition time, the better the accuracy, where the peak position can be located to an accuracy that is substantially better than the actual signal width. On the other hand the accuracy can be limited by systematic errors. To check for this, the data needs to be compared to the simulation: the deviation of angles of the ellipticity peaks over the full data set then gives an upper limit to the accuracy.

7.4 Simulation

For this experiment more sophisticated semi classical simulations were performed by Mathias Smolarski [49]. In addition to the ionization probability also an initial momentum distribution of photoelectrons from helium atoms were modeled with the ADK formalism [82].

The inclusion of an initial momentum distribution is computationally only feasible by using a Monte-Carlo algorithm where the final momentum distribution is obtained by calculating an ensemble of traces in momentum space. The electrons were propagated classically in the calculated pulse field. The starting times were chosen in a range of 10 fs centered on the field maximum in steps of 0.01 fs. For each starting time, 400 traces were calculated and the initial velocity was chosen to match the momentum distribution predicted by the ADK theory. To match the experimental conditions, an initial thermal distribution of 2.8 K in the jet was included. This value was extracted from calibration measurements with linear light, where solely the temperature gives the spread perpendicular to the polarization direction.

For an estimate of the width of the electron wave packet that ultimately limits angular and thus temporal resolution, TDSE calculations were performed by Harm Muller [49]. The time-dependent Schrödinger equation for a helium atom in a circularly polarized IR field was integrated for a peak intensity of 100 TW/cm^2, while putting a pulsed source of electrons on the nucleus. This source term was designed to emit electrons in the same direction as for IR-induced tunneling by populating a suitably chosen superposition of the threshold p-wave and d-wave. It could be switched on and off rapidly without non-adiabatic effects on the wave function inside the atomic potential well. This is different to any other attempts to quickly switch on a tunneling current with a short high-energy electric field wave packet. With this approach the narrowest angular spread of the free electrons was determined to be 35 degrees FWHM, corresponding to a resolution of ≈230 as at central wavelength of the laser pulse of 725 nm.

7.5 Data analysis

7.5.1 Temporal accuracy

To evaluate the temporal accuracy, the angular positions of the ellipticity peaks on the measured momentum distributions were determined and compared to the semi-classical simulation. The results are shown in Figure 7-3.

For each of the 36 subsets of the CEP ramp, the radially integrated angular distribution is fitted with a double Gaussian function to extract the angle of the peaks. Figure 7-3 a and b shows this fit for two different CEP settings. The rms error given from the fitting procedure of the maxima of each of the two ionization peaks (see error bars) gives an upper limit to the accuracy due to the statistical quality of the data. This value could be improved with longer measurements. The calculated error from the peak fitting procedure is less than 3.5 degrees or ≈24 as. The same procedure was applied to the simulated momentum distributions.

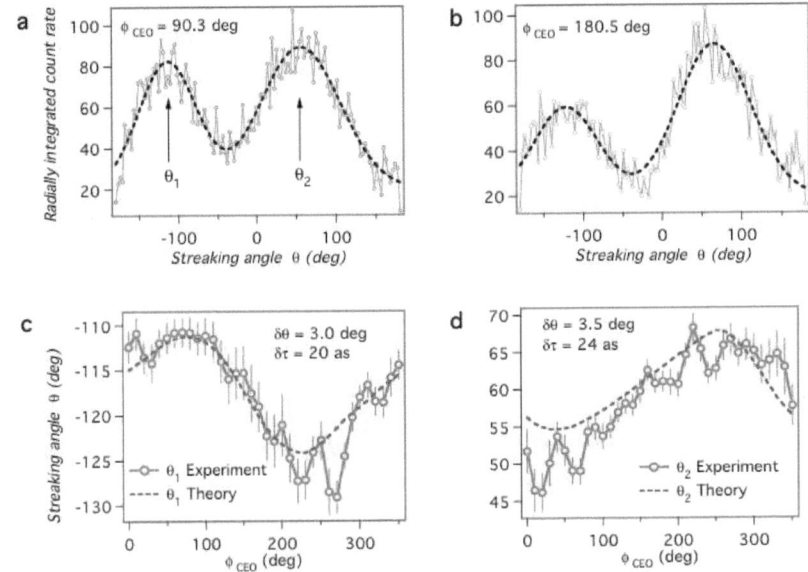

Figure 7-3 CEP dependence of the ionization angle in helium using attosecond angular streaking: a, b: Radially integrated ion momentum distributions for two values of the CEP, the ellipticity peaks are fitted with double Gaussians to extract $\theta_{1,2}$ c, d: The angular position of the two peaks θ_1 and θ_2 as a function of the CEP with simulations indicated as a dashed line.

The extracted values for the peak positions were then compared between measurement and simulation. This comparison yields an rms error between the two dataset for both oscillation peaks of also 24 as. This means that for the available dataset, the semi-classical model describes this ionization and streaking process with sufficient accuracy.

7.5.2 Temporal resolution

Several independent experimental uncertainties limit the resolution. To find the overall experimental uncertainty in the angular peak position, all errors are converted to an angular spread. The examined uncertainties all stem from different parts in the experiments and are thus assumed to be unrelated. Furthermore all errors should be statistically distributed in a Gaussian distribution. Under these assumptions, the uncertainties can be added as follows:

$$\sigma = \sqrt{\sum_{i=1}^{N} \sigma_i^2}$$

In this experiment, the detector resolution was around 3 degrees, corresponding to ≈20 as. The CEP fluctuations measured by an independent f-to-2f interferometer [83] had a width of 6 degrees or ≈40 as. The temperature of the helium target of 2.8 K caused a radial and angular broadening of the momentum distribution adding an uncertainty of ~8 degrees or 52 as.

The resulting uncertainty can then be calculated to ~10 degrees or ~ 60 as:

More severe than the experimental uncertainty is the inherent uncertainty caused by the initial momentum distribution of the electron wave packet.

The calculated resolution in this case from ADK tunneling theory results in ≈160 as. A full quantum mechanical model based on the time-dependent Schrödinger equation described in 7.4 would predict a maximum resolution of ≈220 as at the used central wavelength of 725 nm.

7.5.3 CEP resolution

The accuracy of the CEP measurement is the same as the temporal accuracy: The data are compared to the simulation; the deviation gives the accuracy, i.e. around 4 degrees. Again within the error limit given by the fitting procedure no systematic errors are detectable. This means that a longer scan would lead to higher accuracy.

The resolution with which the CEP can be determined is different to the temporal resolution of AAS.

The resolution is the ability to assign a CEP value to an ion distribution measured at a fixed CEP. In elliptically polarized light, where the relation between peak angles and CEP is a complicated function depending on the ellipticity, this function needs to be known. The achievable resolution is then a combination of the error bar in the peak extraction and the differnce in peak angle between adjacent CEP values. This becomes clear from Figure 7-3c: Peak angles are extracted in steps of 10 degrees. In the CEP range between 30 and 90 degrees, the angular position of the peak is almost constant and the change is much smaller than the error bar, leading to a high insecurity in the determination of the CEP. For CEP values between ~130 and ~200 degrees, where the envelope points more towards the major axis of the ellipse the angular position of the peak depends much more strongly on the CEP, thus improving the resolution.

In the experiment shown here, the full range over which the ellipticity peak was calculated to move due to the change in CEP was around of 10 degrees. For higher ellipticities, i.e. more circular light, the influence of the CEP on the peak position increases over the influence of the polarization direction, leading to a higher range in the angular position and thus to higher resolution. Best CEP resolution would thus be achieved for perfectly circular light.

Chapter 8

Absolute time: Tunneling ionization dynamics

8.1 Introduction

The 'tunneling delay time' τ_D was introduced in chapter 2.5. It is defined as a possible delay between a driving electric pulse field that causes an oscillation in the height of the potential barrier of an atom and the resulting rate of the particles that tunnel through this barrier.

The question if the ionization rate adjusts adiabatically to the oscillation of the driving field has been addressed in experiments with linearly polarized light [11] and yielded an upper limit for the duration of a combined excitation and tunneling event of 320 as. In linearly polarized light, the ionization rate can be resolved in time with pump probe techniques. The timing of the electric field oscillation on the other hand cannot be accessed directly, so no clear start of the tunneling process can be measured. In circular or elliptical polarization however, time and space are intimately connected through the rotation of the electric field vector in the polarization plane, so that both the field and the corresponding ionization rate can be measured and their timing compared.

8.1.1 Relative and absolute measurement of the tunneling delay time

The key idea to measure the tunneling delay time τ_D is to find a temporal feature that can be identified in a measurement of the electric field as well as in the temporal evolution of the tunneling rate. Such a feature is given by the ellipticity peaks on the temporal pulse field described in chapter 3.3. These ellipticity peaks correspond to the major axis of the polarization ellipse in space.

Figure 8-1 sketches the idea: In the first picture, the polarization ellipse of an elliptically polarized pulse is shown. Through ionization, the polarization ellipse is translated to an 'ionization rate ellipse' depicted in the second picture. The three ellipses slightly offset in angle indicate the angular dependence on the CEP as discussed in chapter 3.3.3.

If the tunneling delay time τ_D is zero, the rate ellipse (averaged over the CEP) should have a major axis aligned exactly with the major axis of the polarization ellipse. A finite tunneling delay time τ_D on the other hand should rotate the rate ellipse compared to the polarization ellipse.

Furthermore, if τ_D depends on the barrier width, the exact angle should depend on the laser intensity since a stronger laser field leads to a narrower barrier and a thus lower tunneling time.

The ionization rate ellipse is experimentally not directly accessible since it is further rotated in the AAS process by a streaking angle $\Delta\theta$ of around 90 degrees as indicated in the third picture of Figure 8-1.

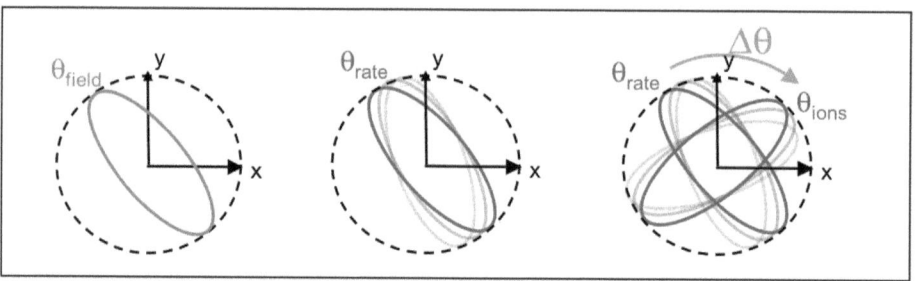

Figure 8-1 Left: the polarization ellipse in red, solid line. Center: the ionization rate ellipse depending on the CEP. On the right: The rate ellipse is streaked in the pulse field and rotated by approximately 90 degrees.

In the momentum distribution, the major axis of the rotated rate ellipse is then given by the ellipticity peaks.

The angle θ_{field} that defines the orientation of the polarization ellipse in space, as shown in Figure 8-1, can be measured by simple polarimetry of the pulse field, yielding the angles of the major and minor axis and their intensity ratio, i.e. the ellipticity of the pulse.

The orientation of the 'ionization rate ellipse' is measured using AAS analogous to the experiment presented in Chapter 7. Again it is not feasible to reconstruct the ionization rate ellipse from the peaks on the ion momentum distribution by reversing the streaking numerically. Instead both the ionization step and the streaking are simulated with the simulation described in Chapter 4 assuming a tunneling delay time of $\tau_D = 0$. The real value of τ_D is then given by the difference between $\Delta\theta_{sim}$ and $\Delta\theta_{meas}$.

8.1.2 Ellipticity dependence

The streaking angle $\Delta\theta$ depends very critically on the precise ellipticity of the pulse (see Figure 8-2). It is therefore crucial to measure and control the ellipticity precisely to achieve a high accuracy for the value of $\Delta\theta$ and thus τ_D.

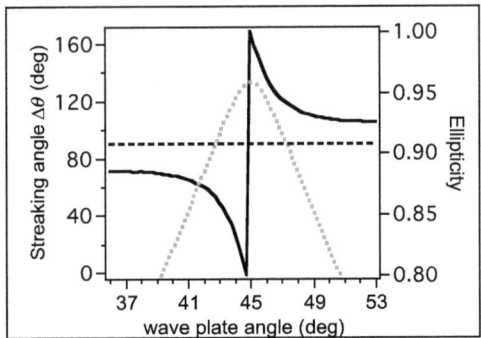

Figure 8-2 shows the strong dependence of the streaking angle on the ellipticity of the pulse. The streaking angle is simulated using the measured pulse parameters and the calculated broadband wave plate used in the experiment: The green dotted line gives the ellipticity as a function of wave plate angle, the black solid line shows the corresponding streaking angle. The streaking angle shows the strongest variation around 45 degrees, where the light is closest to circular.

Furthermore the spatial direction of the major and minor axis and the ellipticity depend also critically on any residual ellipticity of the beam before the quarter wave plate. In the experiment it turned out that it was not possible to avoid all residual ellipticity, since not even the beam directly after the laser is perfectly linearly polarized. Even if the beam is carefully kept at the same height throughout the setup a residual ellipticity of around 0.02 seemed to be the best achievable without further filtering.

8.1.3 CEP dependence

As indicated in Figure 8-1, there is one fundamental difference between the measurement of θ_{field} and θ_{ions}:

The polarization measurement integrates the pulse temporally; this means that there is no information about the CEP of the pulse. Parseval's theorem (see formula below) states that a temporally integrated measurement of the field in time corresponds to a measurement of the spectral intensity (i.e. the intensity in all spectral components separately), so the information about the relative phase of the spectral components and therefore in particular the CEP is lost:

$$\int_{-\infty}^{\infty}|x(t)|^2 dt = \int_{-\infty}^{\infty}|X(f)|^2 df,$$

where $|X(f)|$ and $x(t)$ are connected by a the Fourier transform: $|X(f)| = F\{x(t)\}$.

In the ionization measurement on the other hand as shown in the measurement in chapter 3.3.3, the angular position of the ellipticity peaks depends on the CEP.

The rather easy solution is to measure CEP averaged ion momentum distributions that yield then only one value of θ_{ions}.

Intuitively this can be easily understood: for a CEP =0 as defined in 3.3.4, the envelope peak points into the direction of the major axis, so that the resulting ellipticity peak points along the angle of the major axis. For any other CEP the ellipticity peaks do not point along the axes of the polarization ellipse. For every CEP=ϕ exists however the symmetrical case of CEP=$-\phi$, where the ellipticity peak is shifted in the opposite direction. This is sketched in Figure 8-3 for CEP values of $\pi/2$ and $-\pi/2$. Averaging the two distribution yields again a distribution that is aligned with the polarization ellipse. The average position of the ellipticity peaks then still points along the direction of the major axis. This means that the sum over a random sample of CEP values shows peaks pointing exactly in the direction of the major axis of ellipse.

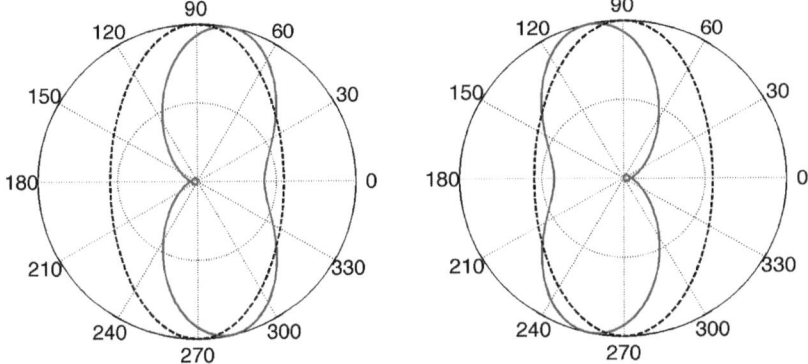

Figure 8-3 Shown are the ionization rate distributions of a few cycle elliptically polarized pulse in the polarization plane for two different CEP values as a solid line. The broken line shows the orientation of the polarization ellipse. On the left, the distribution is shown for a CEP of $\pi/2$, on the right for a CEP of $-\pi/2$. If the two distributions are averaged, the resulting maxima are aligned with the polarization ellipse.

Experimentally, a free running laser provides such a random sample of CEP values.

8.2 Experimental details

8.2.1 Setup

The setup is depicted in Figure 8-4. For the polarization measurement the elliptically polarized pulses were sent through a broadband polarization filter. The energy of the transmitted part of the pulse was then measured as a function of polarizer orientation with a calibrated power meter.

The spatial orientation of the polarization ellipse is expected to vary with frequency, as described in chapter 3.3.3. This means that a uniform spectral response of both the polarizer and the power meter is essential since the power meter spectrally integrates. The power meter and polarization filter[7] were chosen accordingly.

---

[7] LPVIS050, Thorlabs GmbH

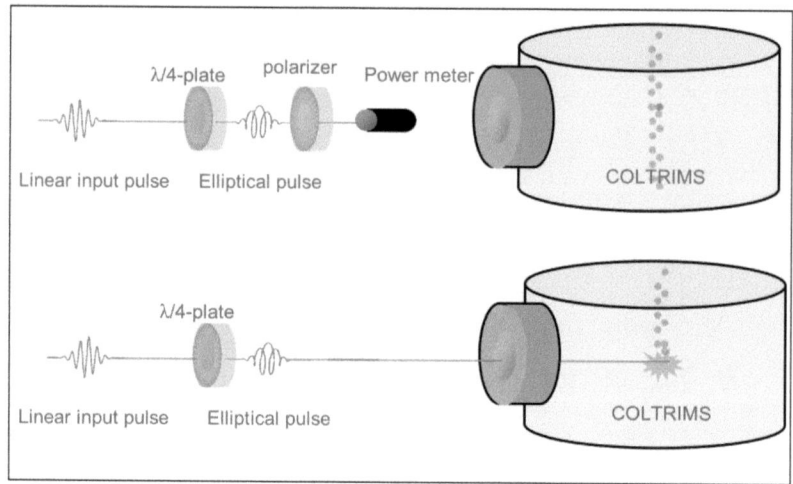

Figure 8-4 The upper panel shows the setup for the polarization measurement, in the lower panel polarizer and power meter are removed to measure the corresponding ion momentum distributions.

Both polarization filter and quarter wave plate were mounted on a computer controlled rotary stage equipped with an absolute position sensor. This setup allowed for precisely repeatable and automatized scans of the orientation of the polarization ellipse and ellipticity with an angular resolution of 0.005 degrees.

After scanning the polarization, the power meter and polarizer could be removed from the setup to let the beam pass into the COLTRIMS for ionization measurements.

8.2.2 Angular calibration

For a measurement of the streaking angle $\Delta\theta$, the relative angle between the coordinate systems of the θ_{field}-measurement and the θ_{ions}-measurement needs to be calibrated.

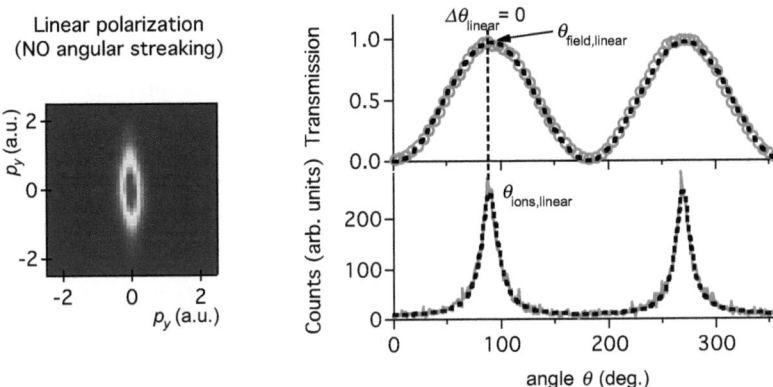

Figure 8-5 Calibration with linear light. On the left the ion momentum distribution in the plane perpendicular to the laser propagation is shown. Vertically polarized light along the y-direction creates a cigar-shaped momentum distributions along the y-direction. On the right the polarization measurement and fit to the radially integrated ion momentum distribution is shown, since this measurement is used to calibrate the coordinate system between ion measurements and polarization measurements the angle $\Delta\theta$ is set to zero.

For an absolute comparison of the measured angles in the polarization measurement and the angular distributions linearly polarized light is used as a reference. In linear light no streaking takes place, i.e. $\Delta\theta = 0$. This means that the angle extracted from the polarization measurement can be set to the value of the angle of the ion momentum peak, thus aligning the coordinate systems for the two measurements. Figure 8-5 shows the calibration measurement with linear light. The upper panel on the right shows the polarization measurement with θ_{field} indicated by an arrow. On the left, the corresponding helium momentum distribution perpendicular to the laser propagation direction is shown. The lower panel on the right then shows the radially integrated ion distribution with θ_{ions} aligned with θ_{field}. All angles shown in the data analysis are calibrated through this procedure.

8.2.3 Measurement procedure

To ensure that the calibration was not affected by alignment, the calibration procedure and measurement were performed in the following sequence:

First the polarization scan was taken for linear light, then the wave plate was placed into the beam path and polarizer scans were recorded for different wave plate angles, leaving the polarizing filter and power meter in place. Then the polarization measurement setup was simply removed without realigning the beam, which could potentially change the angle of incidence on the wave plate. Ion data was taken for the same sequence of wave plate angles, using the computer controlled rotary stage to reliably repeat the wave plate scan that was taken for the polarization measurement. Finally the wave plate was removed to measure the ion distribution generated by linearly polarized light.

8.3 Measurement of the tunneling delay time in an ellipticity scan

8.3.1 Polarization scan

For each angle of the wave plate, a full 360-degree scan of the polarizing filter was performed. The angular data was recorded together with the intensity data from the power meter. A typical scan is shown in Figure 8-6 on the right in the upper panel. The two peaks determine the direction of the polarization ellipse and the ratio between maxima and minima of the curve determines the ellipticity.

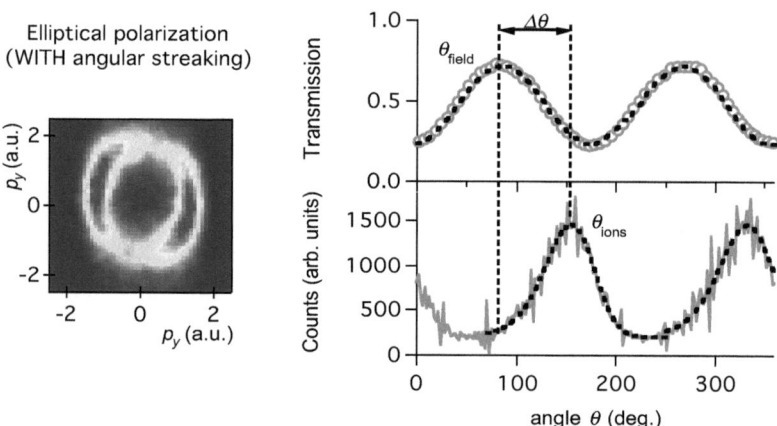

Figure 8-6 On the left, a typical ion momentum distribution projected onto the polarization plane is shown. On the right as in Figure 8-5, in the upper panel the polarizer scan is shown with the angular orientation of the polarization ellipse, θ_{field}. The lower panel shows the corresponding angular distribution of the ions with the maximum θ_{ions} shifted by the streaking angle $\Delta\theta$.

These curves were normalized to a maximum transmission intensity of one and fitted with the following function:

$$I_{meas} = \varepsilon^2 \cdot \sin(\theta - \theta_{field})^2 + \cos(\theta - \theta_{field})^2$$

where ε is the ellipticity and θ_{field} is the peak position. In the experiment, the wave plate angle was scanned over ~50 degrees in ~1 degree steps.

At 45 degrees, where equal amounts of light are coupled into the two crystal axes, theoretically the pulses could become perfectly circular, so that no ellipticity peaks could be defined any more. As explained in chapter 4.2.2, in reality the maximum achievable ellipticity is around 0.95 due to the imperfect wavelength dispersion in the quarter wave plate. The position of maximum ellipticity used to find the absolute angular reference of the wave plate angle in the experiment.

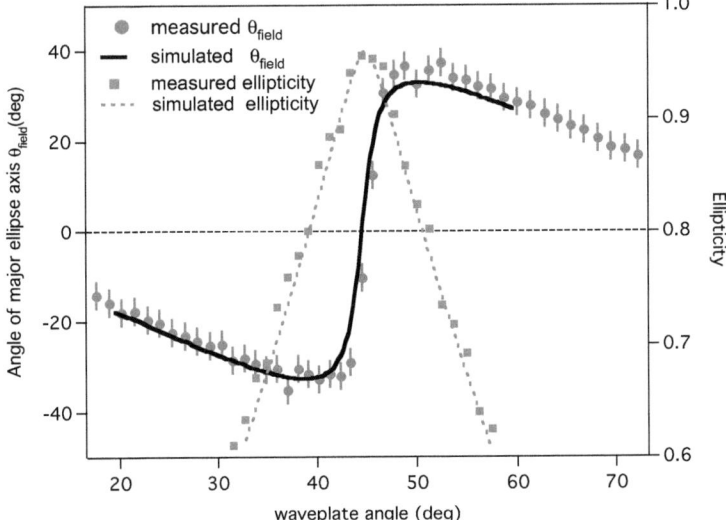

Figure 8-7 Measurement of θ_{field}. Scan of the wave plate angle and thus the ellipticity. Shown as dots is the measured orientation of the polarization ellipse, the solid line shows the simulated orientation. Green squares show the ellipticity extracted from the measurement and the dotted line gives the corresponding simulated ellipticity curve.

The results are shown in Figure 8-7. Each red point corresponds to one setting of the wave plate; the value of θ_{field} is indicated on the left axis. Between ~30 and ~60 degrees of wave plate angle also the corresponding ellipticity is shown for every setting of the wave plate as a green dotted line. Ellipticity values are indicated on the axis on the right. The error bars are given by the fitting procedure described in chapter 8.4.3.

8.3.2 Ion measurement

For the measurement of θ_{ion}, the wave plate was again scanned in 1 degree steps, spanning from 35 to 55 degrees. This range is smaller than for the measurement of θ_{field}, because for wave plate angles outside this interval the peaks on the momentum distributions are not well resolved any more; the streaking in radial direction becomes too low. The angular position of the ellipticity peaks, θ_{ion}, were calculated by first radially integrating the angular momentum data and then fitting with a double peak Gaussian distribution, in the same procedure as used in the analysis in 7.5.

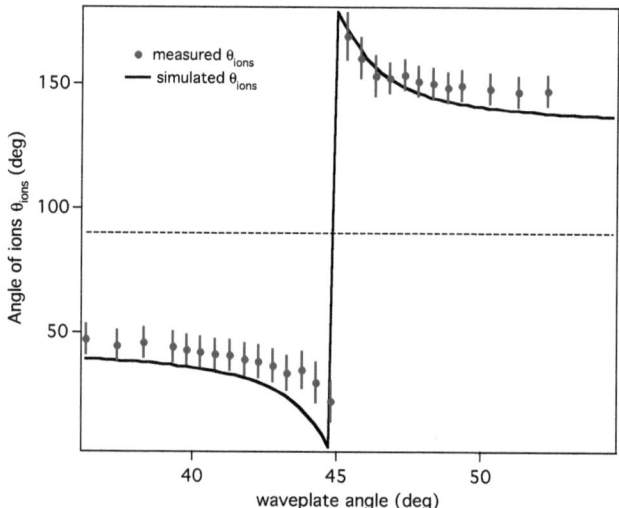

Figure 8-8 Measurement of θ_{ions} Scan of the wave plate angle and thus the ellipticity. Shown as dots is the measured orientation of the ellipticity peaks on the ion momentum distribution, the dashed line shows the simulated orientation.

A typical ion momentum distribution projected on the polarization plane is shown in Figure 8-6 on the left. The lower panel on the right shows the radially integrated momentum distribution with the fitted peaks indicated by a broken line. Figure 8-8 shows the results for the scan of the wave plate angles and thus the ellipticity. Blue dots with error bars give the measured value of θ_{ions}. The solid line shows the corresponding simulation. The systematic angular offset between data and experiment is due to the Coulomb force that was not taken into account in the simulation.

8.4 Data analysis

8.4.1 Simulations

Simulations were performed as explained in Chapter 4 and extended to simulate the scan of the angular orientation of the wave plate. The measured residual ellipticity of the input pulse was also taken into account. Since the momentum distribution can be calculated only for a fixed CEP, distributions were calculated for equally spaced CEP values covering 2π and then averaged.

Peak fitting was used to extract the angles of the polarization ellipse $\theta_{field,sim}$ as well as the angles of the ellipticity peaks of the calculated momentum distributions $\theta_{ion,sim}$. From the simulations, the orientation of the polarization ellipse can be easily calculated by integration over the temporal pulse field.

The results of the simulations are shown as solid lines in Figure 8-7 and Figure 8-8, respectively.

8.4.2 Streaking angle and tunneling delay time

Finally from these two datasets the streaking angle $\Delta\theta$ can be calculated for both experiment and simulation.

Figure 8-9 shows the results: The green dotted line gives the ellipticity. The values of the streaking angle extracted from the data is given by green dots with error bars combined from the error bars derived of the measurements of θ_{field} and θ_{ions}. The broken line is the simulated streaking angle $\Delta\theta_{sim}$. In the simulations described in Chapter 4, the Coulomb force between ion and electron during the streaking was neglected.

The effect of the Coulomb potential was then calculated in a separate simulation by Harm Muller [58]. It depends weakly on the ellipticity. The calculated offset angles were added to the original simulation, the resulting curve is shown as a black solid line in Figure 8-9. The simulations agree with the measured streaking angle for almost all values of ellipticity. This means that within the accuracy of this measurement, no offset between the simulation and the measurement could be found.

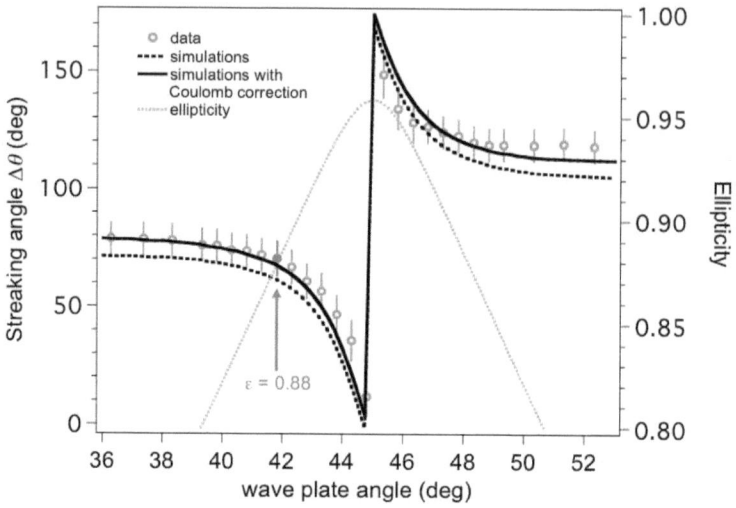

Figure 8-9 The streaking angle $\Delta\theta$ from the measurement is shown as dots with error bars derived from the individual measurements of θ_{field} and θ_{ions}. The black broken line shows the corresponding simulation. The calculated values were corrected by the effect of the coulomb potential yielding the values shown as a black solid line. The green dotted line gives the corresponding ellipticity indicated on the axis on the right.

8.4.3 Statistical error limit

The polarization and the ion distribution curves are fitted by an iterative nonlinear fitting procedure, using target functions $f(x,\beta)$, where β represents a set of parameters to be optimized. The 'best values' of the angular direction θ_{field} and θ_{ions} together with their standard deviations σ_{field} and σ_{ions} are estimated with a Levenberg-Marquardt algorithm that is based on the minimization of χ^2:

$$\chi^2(\beta) = \sum_i \left(\frac{f(x_i,\beta) - y_i}{\sigma_i} \right)^2$$

where $f(x_i,\beta)$ is the fitted value for a given measured data point (x_i, y_i) and σ_i is the standard deviation for y_i, i.e. the weighting given by a known experimental measurement error. In this case the curve fitting is done without weighting (i.e. σ_i all set to 1). The target function for the polarization data was a $\sin^2$-function and for the ion data a double Gaussian function. This fitting procedure then yields the value of $\theta_{field} \pm \sigma_{field}$ and $\theta_{ions} \pm \sigma_{ions}$ for any ellipticity.

The error bars ($\sigma_{\Delta\theta}$) for the values of $\Delta\theta = \theta_{ions} - \theta_{field}$ in Figure 8-9 result from the Gaussian propagation of the individual uncertainties on θ_{ions} and θ_{field}:

$$\sigma_{\Delta\theta} = \sqrt{\sigma_{ions}^2 + \sigma_{field}^2}$$

The same approach is used for the error estimation in Figure 8-10: for every intensity, the error bars are calculated as described above resulting in individual standard deviations ranging between 13 and 20 as.

Motivated by our quantum mechanical calculations that predict an instantaneous tunneling delay time, an upper limit of $\Delta\tau_D$ was determined from the intensity-averaged weighted mean. The weighted mean is calculated as:

$$\Delta t_{D,mean} = \frac{\sum_i \Delta t_D^i / \sigma_{\Delta t_D^i}^2}{\sum_i 1/\sigma_{\Delta t_D^i}^2}$$

and the standard deviation of the weighted mean as:

$$\sigma_{\Delta t_{D,mean}} = \sqrt{\frac{1}{\sum_i 1/\sigma_{\Delta t_D^i}^2}}$$

resulting in an intensity-averaged weighted mean $\Delta\tau_{D,mean}$ of less than 1 degree (corresponding to 6.0 as) with a standard deviation of the weighted mean of 5.6 as. This would lead to an upper limit of a possible tunneling delay time of less than 12 as.

A more conservative upper limit can be estimated by using the data point with the largest deviation from zero tunneling delay time, which occurs at the lowest intensity. Using this data point with 13.9 as and a standard deviation of 19.7 as obtained from the fitting procedure results in a conservative upper limit of 34 as.

8.5 Intensity dependence of the tunneling delay time

The tunneling time suggested by Buttiker as well as the imaginary tunneling time derived from the Keldysh parameter γ in [45] depend on the Keldysh parameter and therefore on the intensity. The laser intensity varies the barrier width; a lower intensity leads to a thicker barrier and therefore a longer tunneling time. For the experimental parameters used in the experiment presented in 8.3 applying this formula would result in 450 as for the barrier interaction time. The time τ_D that was measured in this experiment was shown to be much smaller, i.e. shorter than 50 as. The question however remains if any change in this time could be measured for different ionization intensities and thus different Keldysh parameters.

This experiment is conceptually even simpler than the measurement of the absolute value of τ_D. A dependence of τ_D on the adiabaticity parameter of Keldysh would simply result in an intensity dependent delay of the ionization rate compared to the ionizing field. Keeping the electric field constant while varying the intensity would then delay the ellipticity peaks for higher tunneling barriers i.e. lower intensities.

Again for this experiment no CEP stabilization is needed since a shift would be the same for all CEP phases.

8.5.1 Intensity scans

Controlling the intensity of a 5 fs pulse is not trivial. Several methods are possible, each with their specific drawbacks.

The easiest way to reduce the intensity of a pulse is transmission through an absorbing or partially reflecting material. The disadvantage is that transmission through any such material introduces dispersion that could, in our case, not be compensated for.

Dielectric beam splitters with reflection coefficients of 37, 50 and 75 percent were tested that can be used in reflection thus avoiding material dispersion. It was however found, that these beam splitters introduce considerable spectral distortions because the reflection coefficient is not constant over the full bandwidth of a 5 fs pulse, and the reflection characteristics also varied between beam splitters. Spectral distortions change important pulse parameters such as cycle duration and pulse length. Another widely used technique to control the intensity for ~30 fs pulses is the combination of a half wave plate and a polarizer. The half wave plate rotates the polarization of linearly polarized light and the polarizer transmits intensity depending on the polarization angle. Broadband wave plates and polarizers are available, but introduce considerable chirp. As this setup can be implemented before compression however, the material can be easily compensated for by the grating or prism compressor after the amplifier. This is not feasible for shorter pulses however, because to create the 5 fs pulse, the intensity that is sent into the filaments needs to be kept constant in order not to change the spectral broadening.

A widely used technique for few cycle pulses is the use of an adjustable iris. Cutting the beam with an iris however changes not only the size of the focus but also its position and Raleigh range. For the experiments presented here this is not critical however, since the target was placed outside the focus of the laser.

More limiting to the range of intensities that could be tested proved the COLTRIMS detection. The ionization rate rises approximately with the fifth power of intensity, so that even small further increases in the intensity would lead to saturation effects on the detectors. Detector saturation leads to uneven detection probabilities at different spots that distort the momentum distribution. The maximum possible count rate was found by checking the symmetry in the momentum distributions. This yielded around 1200 counts overall per second limiting the maximum intensity to $3.3 \cdot 10^{14} W/cm^2$. The corresponding Keldysh parameter is then 1.23, so that the threshold of $\gamma = 1$ could not be crossed. In principle there is no limit to go to lower count rates except impractically long acquisition times.

8.5.2 Experiment

For this experiment the ellipticity was kept constant at a value of $\varepsilon = 0.88$. This point is indicated by the filled, red dot in Figure 8-9, choosing a point at the curve slightly off the highest ellipticity. In general the streaking process is most uniform and gives the best angular resolution around values of high ellipticities, i.e. a wave plate angle of ~ 45 degrees. On the other hand, the streaking angle $\Delta\theta$ becomes very sensitive to the exact value of the ellipticity approaching perfectly circular light. In addition, for higher ellipticities the contrast in the polarization measurement and in the ion data becomes lower which leads to higher statistical errors in the peak fitting procedure and a larger influence of the residual ellipticity of the original pulse.

Count rates were varied between 100 Hz and 1200 Hz, yielding intensities of $2.3 \cdot 10^{14} W/cm^2$ to $3.3 \cdot 10^{14} W/cm^2$.

Figure 8-10 shows the result of this measurement: within the error bars, no intensity dependence could be found. This curve is also gives absolute streaking angles (y-scale on the left) and converts them to time delays on the basis of a central wavelength corresponding to 725 nm (y-scale on the right). To find these absolute values, the same procedure as in the measurement of the ellipticity dependence of the streaking angle is applied. The angular orientation of the polarization ellipse θ_{field} does not vary with intensity, also the ADK-rate based simulations that assume no tunneling delay time τ_D yield intensity independent streaking angles $\theta_{ion,sim}$. The streaking angles extracted from the simulation $\Delta\theta_{sim}$ were then corrected for the Coulomb force that was neglected in the calculation of the streaking process. Also the Coulomb correction shows no significant intensity dependence and thus generates a constant offset of the streaking angle.

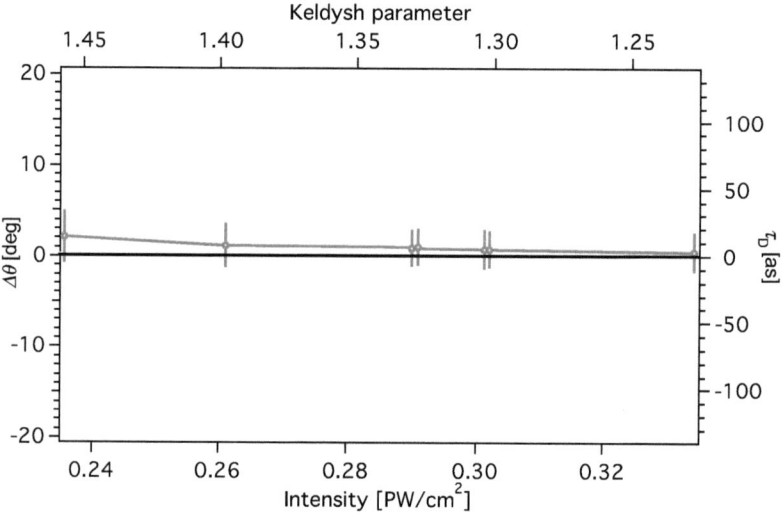

Figure 8-10 Measured tunneling time delay $\Delta\tau_D$ (data points) for a fixed ellipticity (i.e. 0.88) as a function of peak intensity and Keldysh parameter. The tunneling time delay is the difference between the measured and the calculated streaking angle assuming instantaneous tunneling. In the calculated streaking angle we also took into account the Coulomb potential (solid line set to zero). The intensity-averaged offset is 6.0 as with a standard deviation of 5.6 as. The two dimensional helium momentum distributions measured at different intensities and the corresponding radially integrated distributions spanning one optical cycle (2.44 fs) are shown on the right. The time axis is calibrated and the calculated streaking time is deducted.

The weighted intensity averaged offset between the simulated and the measured streaking angle amounts to 6 as, with a standard deviation of 6 as, taking into account the individual data errors. (See also the description of the error treatment procedure in 8.4.3). This yields an upper limit for the tunneling delay time τ_D of 12 as.

8.6 Conclusion

In this chapter the tunneling time delay τ_D was measured for the first time using AAS. A time marker was introduced that can be measured in the electric field as well as in the corresponding ionization rate with a precision of much less than a cycle of the central wavelength of the pulse.

To determine τ_D, measurements were compared to a simulation that assumes that the ionization rate follows the electric field adiabatically. A finite delay between the field and the resulting ionization rate would then appear as an angular offset between the streaking angle extracted from the measurement, $\Delta\theta_{meas}$ and the streaking angle found from the simulation $\Delta\theta_{sim}$.

First, measurements were taken for different ellipticities, at one fixed value of the laser intensity of $2.3 \cdot 10^{14} W/cm^2$, i.e. a Keldysh parameter of 1.46. τ_D is not expected to depend on the ellipticity in first order because the fundamental parameters, the width of the tunneling barrier and the laser oscillation frequency are constant.

In the second experiment, the ellipticity was then kept constant at $\varepsilon = 0.88$, and the intensity was varied between $2.3 \cdot 10^{14} W/cm^2$ and $3.3 \cdot 10^{14} W/cm^2$, corresponding to a variation in the Keldysh parameter of 1.45 to 1.17. τ_D was expected to vary with the barrier width. In both measurements it was found that τ_D is zero within the experimental error limit of 12 as. This result is consistent with quantum mechanical calculations based on solving the time dependent Schrödinger equation [58].

Chapter 9

Summary & Outlook

In this thesis the technique 'attosecond angular streaking' (AAS) was introduced. It is based on ionization and streaking with high-intensity few-cycle pulses with a central wavelength in the near infrared and circular polarization.

Originally AAS was proposed as a tool to measure the absolute value of the carrier envelope offset phase (CEP) that determines the exact shape of the electric field in an ultra short pulse and thus has a strong influence in any strong-field process such as for example high harmonic generation.

In this thesis it was shown that AAS can be applied much more generally as a streaking technique to resolve attosecond time dynamics in strong field processes: The electric field vector of the pulse that rotates in the polarization plane creates a unique relation between time and momentum within one cycle of the carrier frequency (2.4 fs) of the pulse.

AAS can be described as a two step process: First, an electron is set free by tunnel ionization through the potential barrier that is suppressed by the strong field of the pulse, and second the streaking, where both ion and electron move under the influence of the pulse field, acquiring a final momentum depending on the time of ionization in the pulse. The distribution of momenta in the polarization plane maps out the ionization rate and thus the temporal evolution of the field strength.

In linearly polarized light, the CEP determines the evolution of the electric field of the pulse in time. AAS on the other hand employs circularly polarized light, where the CEP determines the angular orientation of the electric field evolution in the polarization plane. To achieve a high resolution of the CEP, ideally the pulse should be as circular as possible. In practice it is not possible to achieve perfectly circular polarization in ultrashort pulses due to their large spectral bandwidths. The residual ellipticity creates a small sub-cycle oscillation of the magnitude of the electric field vector while it traces the polarization ellipse. This sub-cycle oscillation masks the effect of the CEP.

In a proof of principle experiment, AAS was applied in a CEP resolved measurement. The CEP was slowly ramped over 2π while momentum distributions of helium ions were recorded with a COLTRIMS apparatus that allows to detect the momentum vector of a particle in three dimensions. The momentum distributions in the polarization plane could then be compared to a simulation. The resolution of the CEP depends, due to the ellipticity, on the value of CEP and thus is not constant over the data set. Compared to other techniques to measure the absolute value of the CEP such as Stereo-ATI where many electrons can be generated and detected per pulse, the measurement in COLTRIMS needs long integration times since it allows typically less than one ionization event per laser shot. A big advantage is however, that the CEP is measured in situ which might be very useful for further experiments: If AAS should be used for absolute time measurements in single events, the CEP needs to be fixed (i.e. stabilized) and its absolute value needs to be known in the interaction region. This is only possible with such a calibration measurement.
It was also shown that careful simulations of the pulse field are necessary but feasible to achieve such high accuracy.

This experiment also demonstrated the time resolution and accuracy that can be achieved with AAS: By comparing the CEP resolved dataset with a semi-classical simulation an overall accuracy of 34 as could be shown. Additional quantum mechanical simulations indicate a temporal resolution that is limited by the spread of the wavepacket of the ionized electron to around 200 as.
In the next step AAS was employed to measure the tunneling delay time τ_D, defined as a possible time delay between the ionizing field and the corresponding tunneling rate. For this experiment the timing of the electric field needed to be measured separately, using a temporal marker provided by the inherent ellipticity in the pulse field. The tunneling delay time τ_D was measured for a range of ellipticities and as a function of intensity. Comparing data and simulations yields a τ_D of 6 as from the intensity dependent measurement, with a standard deviation of 6 as. This means, that within the limit of 12 as given by the experimental error, τ_D is zero. This result is in accordance with simulations that solve the time dependent Schrödinger equation.
AAS might be an approach to study another tunneling time, τ_{BL}, that was introduced by Büttiker and Landauer as discussed in chapter 2. In their model, an oscillating barrier is used to probe the interaction time of a particle with the tunneling barrier. The barrier

oscillation leads to sidebands in the energy of the transmitted particles as they pick up or loose oscillation quanta in the tunneling process. The relative strength of these sidebands allows to calculate the barrier interaction time. The Büttiker-Landauer model used a barrier solely varying in height, while in AAS the field vector changes the angular dependence of the barrier. Efforts are under way to transfer the model of the oscillating barrier to the case of a spatially changing barrier.

Attosecond pulses alone with their high energy photons and currently low intensity do not induce strong field processes. They have however been successfully combined as triggers or temporal references with strong field few-cycle pulses in a streaking experiment similar to AAS, energy streaking. AAS could in principle also be combined with an attosecond pulse in a similar way. Compared to energy streaking that uses linearly polarized light it has the advantage that the measurement span is a full cycle of the streaking field while in energy streaking the mapping is only unique within a quarter cycle between a maximum and a minimum of the field and the mapping function is only linear around its zero crossing. AAS on the other hand is limited by the initial angular spread in the streaked particle, prohibiting to study processes that are angularly isotropic such as ATI.

All experiments presented in this thesis used helium as it has a perfectly symmetric s-orbital as ground state. On this model system it could be shown that the pulse field can be simulated very accurately which is crucial to achieve high accuracy in a time measurement. As a next step AAS can now be applied to different atomic species or molecules to study complex processes such as electronic correlations for example in double ionization.

References

[1] A. H. Zewail, "Femtochemistry: atomic-scale dynamics of chemical bond," *J. Phys. Chem. A,* vol. 104, pp. 5660-5694, 2000.

[2] R. J. Levis, G. M. Menkir, and H. Rabitz, "Selective Bond Dissociation and Rearrangement with Optimally Tailored, Strong-Field Laser Pulses " *Science,* vol. 292, pp. 709-713, 2001.

[3] J. H. Posthumus, "The dynamics of small molecules in intense laser fields," *Rep. Prog. Phys.,* vol. 67, pp. 623-665, 2004.

[4] A. E. Kaplan, "Essay: The long and the short of it... Time: How much of the cosmological timescale do we control and use? ," *Nature,* vol. 431, p. 633, 2004.

[5] T. Brabec and F. Krausz, "Intense few-cycle laser fields: Frontiers of nonlinear optics," *Rev. Mod. Phys.,* vol. 72, pp. 545-591, 2000.

[6] G. Steinmeyer, D. H. Sutter, L. Gallmann, N. Matuschek, and U. Keller, "Frontiers in Ultrashort Pulse Generation: Pushing the Limits in Linear and Nonlinear Optics," *Science,* vol. 286, pp. 1507-1512, 1999.

[7] E. Goulielmakis, M. Schultze, M. Hofstetter, V. S. Yakovlev, J. Gagnon, M. Uiberacker, A. L. Aquila, E. M. Gullikson, D. T. Attwood, R. Kienberger, F. Krausz, and U. Kleineberg, "Single-Cycle Nonlinear Optics," *Science,* vol. 320, pp. 1614-1617, 2008.

[8] G. Sansone, E. Benedetti, F. Calegari, C. Vozzi, L. Avaldi, R. Flammini, L. Poletto, P. Villoresi, C. Altucci, R. Velotta, S. Stagira, S. De Silvestri, and M. Nisoli, "Isolated single-cycle attosecond pulses," *Science,* vol. 314, pp. 443-446, 20.10.2006 2006.

[9] E. Goulielmakis, M. Uiberacker, R. Kienberger, A. Baltuska, V. Yakovlev, A. Scrinzi, T. Westerwalbesloh, U. Kleineberg, U. Heinzmann, M. Drescher, and F. Krausz, "Direct Measurement of Light Waves," *Science,* vol. 305, pp. 1267-1269, 2004.

[10] M. Drescher, M. Hentschel, R. Kienberger, M. Uiberacker, V. Yakovlev, A. Scrinzi, T. Westerwalbesloh, U. Kleineberg, U. Heinzmann, and F. Krausz, "Time-resolved atomic inner-shell spectroscopy," *Nature,* vol. 419, pp. 803-807, 2002.

[11] M. Uiberacker, T. Uphues, M. Schultze, A. J. Verhoef, V. Yakovlev, M. F. Kling, J. Rauschenberger, N. M. Kabachnik, H. Schröder, M. Lezius, K. L. Kompa, H.-G. Muller, M. J. J. Vrakking, S. Hendel, U. Kleineberg, U. Heinzmann, M. Drescher, and F. Krausz, "Attosecond real-time observation of electron tunnelling in atoms," *Nature,* vol. 446, pp. 627-632, 2007.

[12] L. A. MacColl, "Note on the Transmission and Reflection of Wave Packets by Potential Barriers," *Physical Review,* vol. 40, p. 621, 1932.

[13] R. Landauer and T. Martin, "Barrier interaction time in tunneling," *Reviews of Modern Physics,* vol. 66, p. 217, 1994.

[14] J. Ruseckas, "Possibility of tunneling time determination," *Physical Review A,* vol. 63, p. 052107, 2001.

[15] E. H. Hauge and J. A. Støvneng, "Tunneling times: a critical review," *Rev. Mod. Phys.,* vol. 61, pp. 917 - 936, 1989.

[16] J. G. Muga, R. S. Mayato, and I. L. Egusquiza, "Introduction," *Lect. Notes Phys.,* vol. 743, pp. 1-30, 2008.

[17] W. Pauli, *Handbuch der Physik, Springer, Berlin,* vol. 23, pp. 1-278, 1926.

[18] E. A. Galapon, "Pauli's theorem and quantum canonical pairs: the consistency of a bounded, self-adjoint time operator canonically conjugate to a Hamiltonian with non-empty point spectrum," *Proc. R. Soc. A,* vol. 458, pp. 451-472, 2002.

[19] E. A. Galapon, R. F. Caballar, and R. Bahague, "Confined quantum time of arrival for the vanishing potential," *Physical Review A (Atomic, Molecular, and Optical Physics),* vol. 72, pp. 062107-17, 2005.

[20] E. A. Galapon, R. F. Caballar, and R. T. B. Jr, "Confined Quantum Time of Arrivals," *Phys. Rev. Lett.,* vol. 93, p. 180406, 2004.

[21] P. Busch, "The Time-Energy Uncertainty Relation," *Lect. Notes Phys.,* vol. 734, pp. 73-105, 2008.

[22] M. Büttiker, "Larmor precession and the traversal time for tunneling," *Phys. Rev. B,* vol. 27, pp. 6178 - 6188 1983.

[23] S. Brouard, R. Sala, and J. G. Muga, "Systematic approach to define and classify quantum transmission and reflection times," *Phys. Rev. A,* vol. 49, pp. 4312-4325, 1994.

[24] E. P. Wigner, "Lower Limit for the Energy Derivative of the Scattering Phase Shift," *Physical Review,* vol. 98, p. 145, 1955.

[25] L. Eisenbud, *PhD. Thesis, Princeton University,* 1948.

[26] M. Büttiker and R. Landauer, "Traversal Time for Tunneling," *Phys. Rev. Lett.,* vol. 49, pp. 1739-1742, 1982.

[27] T. E. Hartmann, "Tunneling of a Wave Packet," *J. Appl. Phys.,* vol. 33, pp. 3427-3433, 1962.

[28] A. Enders and G. Nimtz, "On superluminal barrier traversal," *J. Phys. I France,* vol. 2, p. 1693, 1992.

[29] A. M. Steinberg, P. G. Kwiat, and R. Y. Chia, "Measurement of the Single-Photon Tunneling Time," *Phys. Rev. Lett.*, vol. 71, pp. 708-711, 1993.

[30] C. Spielmann, R. Szipöcs, A. Stingl, and F. Krausz, "Tunneling of Optical Pulses through Photonic Band Gaps," *Phys. Rev. Lett.*, vol. 73, pp. 2308-2311, 1994.

[31] P. Balcou and L. Dutriaux, "Dual Optical Tunneling Times in Frustrated Total Internal Reflection," *Physical Review Letters*, vol. 78, p. 851, 1997.

[32] M. Mojahedi, E. Schamiloglu, F. Hegeler, and K. J. Malloy, "Time-domain detection of superluminal group velocity for single microwave pulses," *Physical Review E*, vol. 62, p. 5758, 2000.

[33] J. J. Carey, J. Zawadzka, D. A. Jaroszynski, and K. Wynne, "Noncausal Time Response in Frustrated Total Internal Reflection?," *Physical Review Letters*, vol. 84, p. 1431, 2000.

[34] S. Longhi, M. Marano, P. Laporta, and M. Belmonte, "Superluminal optical pulse propagation at 1.5 mum in periodic fiber Bragg gratings," *Physical Review E*, vol. 64, p. 055602, 2001.

[35] A. Hache and L. Poirier, "Long-range superluminal pulse propagation in a coaxial photonic crystal," *Applied Physics Letters*, vol. 80, pp. 518-520, 2002.

[36] A. Sommerfeld, "Ein Einwand gegen die Relativtheorie der Elektrodynamik und seine Beseitigung. An der Diskussion beteiligten sich W. Wien, Braun, Voigt, Des Coudres. (Eingegangen 30. Oktober 1907.)," *Physikalische Zeitschrift*, vol. 8, pp. 841-842, 1907.

[37] L. Brillouin, "Über die Fortpflanzung des Lichtes in dispergierenden Medien," *Ann. Phys.*, vol. 349, pp. 203-240, 1914.

[38] M. Büttiker and H. Thomas, "Front propagation in evanescent media," *Ann. Phys.*, vol. 7, pp. 602-617, 1998.

[39] H. G. Winful, "The meaning of group delay in barrier tunnelling: a re-examination of superluminal group velocities," *New J. Phys.*, vol. 8, 2006.

[40] R. Landauer, "Barrier traversal time," *Nature*, vol. 341, pp. 567-568, 1989.

[41] P. Guéret, E. Marclay, and H. Meier, "Investigation of possible dynamic polarization effects on the transmission probability of n-GaAs/AlxGa1-xAs/n-GaAs tunnel barriers," *Solid State Communications*, vol. 68, pp. 977-979, 1988.

[42] P. Guéret, E. Marclay, and H. Meier, "Experimental observation of the dynamical image potential in extremely low GaAs/AlxGa1☐xAs/GaAs tunnel barriers," *Appl. Phys. Lett.*, vol. 53, p. 1617, 1988.

[43] D. Esteve, J. M. Martinis, C. Urbina, E. Turlot, M. H. Devoret, H. Grabert, and S. Linkwitz, "Observation of the Temporal Decoupling Effect on the Macroscopic Quantum Tunneling of a Josephson Junction," *Physica Scripta,* vol. T29, pp. 121-124, 1989.

[44] L. V. Keldysh, "Ionization in the field of a strong electromagnetic wave," *Sov. Phys. JETP,* vol. 20, p. 1307, 1965.

[45] G. L. Yudin and M. Y. Ivanov, "Nonadiabatic tunnel ionization: Looking inside a laser cycle," *Phys. Rev. A,* vol. 64, pp. 013409, 1-4, 6 June 2001 2001.

[46] G. G. Paulus, F. Zacher, H. Walther, A. Lohr, W. Becker, and M. Kleber, "Above-Threshold Ionization by an Elliptically Polarized Field: Quantum Tunneling Interferences and Classical Dodging," *Physical Review Letters,* vol. 80, p. 484, 1998.

[47] R. Kienberger, M. Hentschel, M. Uiberacker, C. Spielmann, M. Kitzler, A. Scrinzi, M. Wieland, T. Westerwalbesloh, U. Kleineberg, U. Heinzmann, M. Drescher, and F. Krausz, "Steering Attosecond Electron Wave Packets with Light," *Science,* vol. 297, pp. 1144 -1148, 2002.

[48] J. Itatani, F. Quéré, G. L. Yudin, M. Y. Ivanov, F. Krausz, and P. B. Corkum, "Attosecond streak camera," *Phys. Rev. Lett.,* vol. 88, 2002.

[49] P. Eckle, M. Smolarski, P. Schlup, J. Biegert, A. Staudte, M. Schöffler, H. G. Muller, R. Dörner, and U. Keller, "Attosecond Angular Streaking," *Nat. Phys. ,* vol. 4, pp. 565-570, 2008.

[50] M. Lewenstein, P. Balcou, M. Y. Ivanov, A. L'Huillier, and P. B. Corkum, "Theory of high-harmonic generation by low-frequency laser fields," *Phys. Rev. A,* vol. 49, pp. 2117-2132, 1994.

[51] H. R. Telle, G. Steinmeyer, A. E. Dunlop, J. Stenger, D. H. Sutter, and U. Keller, "Carrier-envelope offset phase control: A novel concept for absolute optical frequency measurement and ultrashort pulse generation," *Appl. Phys. B,* vol. 69, pp. 327-332, 1999.

[52] D. J. Jones, S. A. Diddams, J. K. Ranka, A. Stentz, R. S. Windeler, J. L. Hall, and S. T. Cundiff, "Carrier-envelope phase control of femtosecond mode-locked lasers and direct optical frequency synthesis," *Science,* vol. 288, pp. 635-639, 2000.

[53] A. Apolonski, A. Poppe, G. Tempea, C. Spielmann, T. Udem, R. Holzwarth, T. W. Hänsch, and F. Krausz, "Controlling the Phase Evolution of Few-Cycle Light Pulses," *Phys. Rev. Lett.,* vol. 85, pp. 740-743, July 24 2000.

[54] M. Kitzler, N. Milosevic, A. Scrinzi, F. Krausz, and T. Brabec, "Quantum theory of attosecond XUV pulse measurement by laser-dressed photoionization," *Phys. Rev. Lett.*, vol. 88, p. 173904, 2002.

[55] R. Kienberger, E. Goulielmakis, M. Uiberacker, A. Baltuska, V. Yakovlev, U. Heinzmann, M. Drescher, and F. Krausz, "Atomic transient recorder," *Nature*, vol. 427, pp. 817-821, 2004.

[56] M. J. Dodge, "Refractive properties of magnesium fluoride," *Appl. Opt.*, vol. 23, pp. 1980-1985, 1985.

[57] G. Szivessy and C. Münster, "Über die Prüfung der Gitteroptik bei aktiven Kristallen " *Ann. Phys.*, vol. 20, p. 703, 1934.

[58] P. Eckle, A. Pfeiffer, C. Cirelli, A. Staudte, R. Dörner, H. G. Muller, M. Büttiker, and U. Keller, "Attosecond ionization and tunneling delay time measurements," *Science*, p. accepted, 2008.

[59] M. V. Ammosov, N. B. Delone, and V. P. Krainov, "Tunnel ionization of complex atoms and of atomic ions in an alternating electromagnetic field," *Sov. Phys. JETP*, vol. 64, pp. 1191-1194, 1986.

[60] J. Ullrich, R. Mooshammer, A. Dorn, R. Dörner, L. P. H. Schmidt, and H. Schmidt-Böcking, "Recoil-ion and electron momentum spectroscopy: reaction-microscopes," *Rep. Prog. Phys.*, vol. 66, pp. 1463 -1545, 2003.

[61] P. F. Moulton, "Spectroscopic and laser characteristics of Ti:Al_2O_3," *J. Opt. Soc. Am. B*, vol. 3, pp. 125-132, 1986.

[62] D. E. Spence, P. N. Kean, and W. Sibbett, "60-fsec pulse generation from a self-mode-locked Ti:sapphire laser," *Opt. Lett.*, vol. 16, pp. 42-44, 1991.

[63] D. H. Sutter, L. Gallmann, N. Matuschek, F. Morier-Genoud, V. Scheuer, G. Angelow, T. Tschudi, G. Steinmeyer, and U. Keller, "Sub-6-fs pulses from a SESAM-assisted Kerr-lens modelocked Ti:sapphire laser: At the frontiers of ultrashort pulse generation," *Appl. Phys. B*, vol. 70, pp. S5-S12, 2000.

[64] U. Morgner, F. X. Kärtner, S. H. Cho, Y. Chen, H. A. Haus, J. G. Fujimoto, E. P. Ippen, V. Scheuer, G. Angelow, and T. Tschudi, "Sub-two-cycle pulses from a Kerr-lens mode-locked Ti:sapphire laser," *Opt. Lett.*, vol. 24, pp. 411-413, 1999.

[65] D. Strickland and G. Mourou, "Compression of amplified chirped optical pulses," *Optics Communications*, vol. 56, pp. 219-221, 1985.

[66] C. P. Hauri, W. Kornelis, F. W. Helbing, A. Heinrich, A. Courairon, A. Mysyrowicz, J. Biegert, and U. Keller, "Generation of intense, carrier-envelope phase-locked few-cycle laser pulses through filamentation," *Appl. Phys. B*, vol. 79, pp. 673-677, 2004.

[67] M. Nisoli, S. Stagira, S. D. Silvestri, O. Svelto, S. Sartania, Z. Cheng, M. Lenzner, C. Spielmann, and F. Krausz, "A novel high-energy pulse compression system: generation of multigigawatt sub-5-fs pulses," *Appl. Phys. B*, vol. 65, pp. 189-196, 1997.

[68] B. Schenkel, J. Biegert, U. Keller, C. Vozzi, M. Nisoli, G. Sansone, S. Stagira, S. D. Silvestri, and O. Svelto, "Generation of 3.8-fs pulses from adaptive compression of a cascaded hollow fiber supercontinuum," *Opt. Lett.*, vol. 28, pp. 1987-1989, 2003.

[69] P. Sprangle, E. Esarey, and J. Krall, "Self-guiding and stability of intense optical beams in gases undergoing ionization," *Physical Review E*, vol. 54, pp. 4211-4232, Oct 1996.

[70] R. Trebino, K. W. DeLong, D. N. Fittinghoff, J. Sweetser, M. A. Krumbügel, and B. Richman, "Measuring ultrashort laser pulses in the time-frequency domain using frequency-resolved optical gating," *Rev. Sci. Instrum.*, vol. 68, pp. 1–19, 1997.

[71] C. Iaconis and I. A. Walmsley, "Self-Referencing Spectral Interferometry for Measuring Ultrashort Optical Pulses," *IEEE J. Quantum Electron.*, vol. 35, pp. 501-509, 1999.

[72] L. Gallmann, D. H. Sutter, N. Matuschek, G. Steinmeyer, U. Keller, C. Iaconis, and I. A. Walmsley, "Characterization of sub-6-fs optical pulses with spectral phase interferometry for direct electric-field reconstruction," *Opt. Lett.*, vol. 24, pp. 1314-1316, 1999.

[73] C. Iaconis and I. A. Walmsley, "Spectral Phase Interferometry for Direct Electric Field Reconstruction of Ultrashort Optical Pulses," *Opt. Lett.*, vol. 23, pp. 792-794, 1998.

[74] W. Kornelis, J. Biegert, J.W.G. Tisch, M. Nisoli, G. Sansone, C. Vozzi, S. De Silvestri, U. Keller, "Single-shot kHz ultrashort pulse characterization using SPIDER," *Opt. Lett.*, vol. 28, pp. 281-283, 2003.

[75] C. Dorrer, N. Belabas, J. P. Likforman, and M. Joffre, "Spectral resolution and sampling issues in Fourier-transform spectral interferometry," *Journal Of The Optical Society Of America B-Optical Physics*, vol. 17, pp. 1795-1802, Oct 2000.

[76] G. Tejeda, B. Maté, J. M. Fernández-Sánchez, and S. Montero, "Temperature and Density Mapping of Supersonic Jet Expansions Using Linear Raman Spectroscopy," *Phys. Rev. Lett.*, vol. 76, pp. 34-37, 1996.

[77] D.R.Miller, "Free Jet Sources," *Atomic and Molecular Beam Methods*, 1988.

[78] O. Jagutzki, J. S. Lapington, L. B. C. Worth, U. Spillmann, V. Mergel, and S. Schmidt-Böcking, "Position sensitive anodes for MCP read-out using induced charge measurement," *Nucl. Instrum. Methods Phys. Res., Sect. A,* vol. 477, pp. 256-251, 2002.

[79] P. Dietrich, F. Krausz, and P. B. Corkum, "Determining the absolute phase of a few-cycle laser pulse," *Opt. Lett.,* vol. 25, pp. 16-18, 2000.

[80] G. G. Paulus, F. Grasborn, H. Walther, P. Villoresi, M. Nisoli, S. Stagira, E. Priori, and S. D. Silvestri, "Absolute-phase phenomena in photoionization with few-cycle laser pulses," *Nature,* vol. 414, pp. 182-4, 2001.

[81] W. Kornelis, J. Biegert, J.W.G. Tisch, M. Nisoli, G. Sansone, C. Vozzi, S. De Silvestri, U. Keller, "Single-shot kilohertz characterization of ultrashort pulses by spectral phase interferometry for direct electric-field reconstruction," *Opt. Lett.,* vol. 28, pp. 281-283, 2003.

[82] N. B. Delone and V. P. Krainov, "Energy and angular electron spectra for the tunnel ionization of atoms by strong low-frequency radiation," *J. Opt. Soc. Am. B,* vol. 8, pp. 1207-1211, 1991 1991.

[83] A. Guandalini, P. Eckle, M. P. Anscombe, P. Schlup, J. Biegert, and U. Keller, "5.1 fs pulses generated by filamentation and carrier envelope phase stability analysis," *J. Phys. B: At. Mol. Opt. Phys.,* vol. 39, pp. S257-S264, 2006

Acknowledgements

An dieser Stelle möchte ich mich bei allen bedanken, die zum Gelingen dieser Doktorarbeit beigetragen haben.

Ganz besonders danke ich Prof. Ursula Keller für die engagierte Betreuung meiner Doktorarbeit, ihre stete Unterstützung, Motivation nicht zuletzt für ihre ansteckende Begeisterung für dieses Vorhaben.

Prof. Reinhard Dörner danke ich für die Übernahme des Koreferats, sein Interesse und seine Begeisterung für unser Projekt.

Ebenfalls danken möchte ich meinen Team Kollegen am COLTRIMS, Adrian Pfeiffer und Dr. Claudio Cirelli und allen anderen Mitgliedern der Gruppe für die gute Atmosphäre, ihre stete Hilfsbereitschaft und das tolle Teamwork.

Besonderer Dank geht auch an meine Familie und Wouter, für ihre Unterstützung und Ermutigung.

Die VDM Verlagsservicegesellschaft sucht für wissenschaftliche Verlage abgeschlossene und herausragende

Dissertationen, Habilitationen, Diplomarbeiten, Master Theses, Magisterarbeiten usw.

für die kostenlose Publikation als Fachbuch.

Sie verfügen über eine Arbeit, die hohen inhaltlichen und formalen Ansprüchen genügt, und haben Interesse an einer honorarvergüteten Publikation?

Dann senden Sie bitte erste Informationen über sich und Ihre Arbeit per Email an *info@vdm-vsg.de*.

Sie erhalten kurzfristig unser Feedback!

VDM Verlagsservicegesellschaft mbH
Dudweiler Landstr. 99
D - 66123 Saarbrücken
www.vdm-vsg.de

Telefon +49 681 3720 174
Fax   +49 681 3720 1749

Die VDM Verlagsservicegesellschaft mbH vertritt

Printed by Books on Demand GmbH, Norderstedt / Germany